THE HUMAN QUESTION

L. Mason Jones

MINERVA PRESS
LONDON
MONTREUX LOS ANGELES SYDNEY

THE HUMAN QUESTION

ISBN 1 86106 819 0

First Published 1998 by
MINERVA PRESS
195 Knightsbridge
London SW7 1RE

Printed in Great Britain for Minerva Press

THE HUMAN QUESTION

I would like to dedicate this book to all those people who through their labours are at least attempting to solve the mystery of human origins, no matter how the answer may astound us.

Author's Note

Throughout this work, where I refer to the 'Cro-Magnon' entity, I mean the anatomically modern human remains dated around 30,000 years old and found at the rock shelter of Cro-Magnon (Les Eyzies Tayac, Dordogne, France). The Cro-Magnon 'type' are found in other areas of the world and are named according to their locations of discovery and dated 30–50,000 years old. Anthropologists are trying to push back the date of anatomically modern humans as far as possible, particularly as it is virtually proven now (genetically) that they were not related to *Neanderthalensis sapiens*, but the older skulls are incomplete and rely on plaster to complete the shape assumed by the anthropologists.

The truth must be found but woe to him who finds it.

Robert Charroux

Interesting Facts

1. Conclusive evidence in bone fossils to prove human ascendancy from the primates has not been found.
2. Eminent anthropologists, such as Richard Leakey admit that it is every palaeontologist's dream to find such links.
3. We can quite easily find bones to construct complete skeletons of creatures for our natural history museums which are *65 million years old.*
4. Bones last for a *very long time* in the ground. Where are the human fossil links?
5. *Neanderthalensis sapiens* seemed like an evolutionary dead end. The highly intelligent Cro-Magnon man, from which we all descend, followed him yet did not seem related or interbred with Neanderthal man, and could not have received his advanced qualities from him.
6. Evolutionary processes need enormous time periods. Frogs, flies, fish and fauna look exactly the same as they did many millions of years ago; Cro-Magnon man was an amazing leap forward in evolutionary development.
7. The human brain is amazingly 'over-endowed' and developed, defying evolutionary and natural selection

processes that demand enormous periods of time, even for the slightest change.

8. Much circumstantial evidence exists that indicates an unusual and continuing interest in human affairs and development from the onset of our recorded history by extraterrestrial intelligence.
9. Alleged worldwide abductions by extraterrestrial beings involves removal and analysis of human genetic material over long periods. This suggests an analysis of human *development*.
10. *Our* geneticists claim to be soon able to 'grow' human tissue, organs, even the heart in laboratories.
11. What would beings only a mere 1,000 years ahead of us be capable of? Some are bound to be more advanced than ourselves.
12. Science admits the possibility of extraterrestrial intelligence.
13. Exobiologists have already admitted our world could have been visited in the past by unearthly intelligence.

Conclusion

The profound possibility exists that alien intelligence could have 'created' Cro-Magnon man in a highly scientific genetic process over 50,000 years ago, possibly providing the 'Broca's convolution' in the brain that gives us our human qualities that the poorly endowed ape, with its small underdeveloped brain compared to humans, does not possess.

Foreword

Travelling on a 'Government Service' passport, L. Mason Jones has spent time in Aden, Bahrain, the Iranian Gulf, Masirah and Cyprus. He has also travelled extensively in Europe. During the last fifteen years he has been involved in the final inspection and quality checks on the fully-built Hawker business jet with British Aerospace, Corporate Jets Inc. and Raytheon USA. He has two sons and a daughter and now resides in Chester. *The Human Question* is one of five manuscripts prepared for publication.

Preface

It is quite easy today to accept the possibility of an extraterrestrial intelligence existing that may have passed through our current state of advancement long ago. There is nothing within these pages that could be considered fantasy-prone or that could never happen, or indeed could never have happened.

For example, prior to our knowledge of meteoric debris, asteroids and so forth, it was an amazing piece of logic to state, 'There are no stones in the sky, therefore stones cannot fall from the sky'.

Hindsight affords us the tendency to be derisive of such historic comments, made continually about the flow of scientific pronouncements since the so-called age of enlightenment began.

Scientists (particularly medical scientists) are now pushing hard against the barriers of ethical restraint and view them as retarding 'necessary experimentation', especially regarding genetics, that they like to cover with the respectable cloak of 'essential to the progress of our ability to save or prolong life'.

On 15th December, 1996, a television programme called *Equinox*, shown an Channel 4, dealt with an amazing experiment that transplanted the living head of a monkey onto another monkey's body and restored the functions of sight and so forth. The comment was made that it could now be possible in a human operation, in the case, for example, of a brain-dead but otherwise functioning body. During the programme it was actually stated that the surgeon may go ahead anyway, ethical restraints or not. It is only a question of time before the barriers are removed, just as the censorship laws have collapsed regarding films and the performing arts, and all sorts of nasty things have crept out of the cellar and into the public eye. It is no use us admitting the possibility of extraterrestrial intelligence, but expecting them all to be less advanced than ourselves. 'They' may have carried out such experiments centuries ago, perhaps even millennia ago, and certain 'life enhancement' experiments of a much more advanced nature could have been carried out *right here on Earth*, with the results of such experiments possibly being the human race.

This possibility, together with the shortcomings of the two main theories for human origins, are looked at within these pages.

Contents

Introduction

Although the generally accepted theories of the evolution of the human species are far from proven, this work does not set out to challenge or dispute them but rather to ask questions of them and in so doing, introduce a speculative and, hopefully thought-provoking, perhaps entertaining, alternative for the appearance of the human species. It has to be said that anthropologists, generally, would view it as a form of professional suicide to voice any misgivings (which at least a few of them must harbour) regarding the widely accepted theory, and because of this, the whole concept has taken on a totally unassailable image of unquestionable truth. To be fair, some anthropologists are quite open about the lack of totally convincing evidence and point out the burning issue among their ranks of discovering the vital fossil links

Even Richard Leaky himself admits that it is every anthropologist's dream to find a complete skeleton of an ancient human ancestor, and that the current human fossil collection is a meagre, fragmented array of isolated teeth, single bones and fragments of skulls.[1] Clearly, in other species and fauna, enormous amounts of time have gone by without any apparent change, yet when one looks at the tremendous and sudden shift forward in the human skull of the Cro-Magnon frontal cranium area, or forehead, to

---

[1] Richard Leaky, *The Origin of Humankind*, Phoenix, 1995.

accommodate the enhanced brain, there is a *distinct and immediate* change from the sloping skulls of the *Homo erectus* being and *Neanderthalensis sapiens*. Not only have the excruciatingly slow evolutionary processes made a huge, sudden, giant leap forward in Cro-Magnon man, he does not even appear to be related to his predecessor, Neanderthal man.

Another anthropologist, Alan Bilsborough, says evidence is scrappy and incomplete regarding chimps and gorillas being related to humans and says almost nothing is known of the lineage from the pongid apes to hominids, and a natural progression to mankind.[2]

To reiterate what I said previously, although profound questions are asked in this work regarding human origins, my intention, when writing *The Human Question*, was not to challenge the theory of human evolution as we know it but simply to highlight its apparent '*unassailability*' and to reflect upon a possible astounding alternative. Many questions are still open regarding human development. Nothing in the work is beyond all possibility and, regarding scientific facts that are stated within the pages, I would add that they have become known due to the efforts of clever and learned people and are simply related by me in their context throughout the work.

However, like any person, I do reserve the right to ask questions and that is all *The Human Questions* is about – no more, no less.

No theory is unassailable until it becomes clearly proven fact, and this applies to all the theories for human origins that exist today. In any well-stocked library, the natural history section abounds with fine, glossy, hardback works

---

[2] Alan Bilsborough, *Human Evolution,* Blackie, 1993.

that are the culmination of much hard work – patiently sifting and cataloguing in hostile climates to produce the material for them – and they contain much valuable data on evolutionary concepts. But it has to be said that the bottom line is when it comes to human origins and a *natural* evolution from the primates, it is *not proven*. The pronouncements in the biblical account of Genesis (written by men) for divine human creation, are seen by many as utterly incredible. There are many people that live good, law-abiding lives who are not inordinately religious and, as such, are able to take a long, objective view of biblical writings, unfettered by any constraints of 'blind faith'. How do they view the concept of a fantastic being coming along, waving a majestic hand and stating, 'Let there be light', and within seven days the entire universe, our sun, solar system, and all the creatures on Earth being created? Many people have no problem at all in accepting it.

But cold, analytical science tells us the process took 15–20 billion years. With regard to the Genesis account, many would say it could *only be* sustained by blind faith. However, for all the theories on human origins, this is the one that most right-thinking beings would *wish* to be true, *if* their logical processes allowed them to accept that it actually occurred that way.

In the stern, God-fearing days of the last century, many people kept their reservations to themselves as convincing scientific logic eroded their beliefs further and further. Then came the 'new Messiah', Charles Darwin, an intelligent thinking man, who was very brave and voiced his theory, knowing the derision and scorn that would be heaped upon him for doing so by the outraged religious factions. For many there was a great sigh of relief. At last, a logical theory to explain all the ancient fossils and the obvious great age of the Earth and totally blow away the last vestiges of beliefs of those already lukewarm in their acceptance of Genesis account of human creation.

They had carried their guilt and misgivings for long enough. Here was someone who would relieve them of it, and many flocked to his banner. The questions is, were they led into a wilderness of doubt and uncertainty? Even Charles Darwin himself was uncertain and stated, 'if the fossil links are not found, then the theory falls down'. We must now ask, *is* the theory falling down?

No theory could seem more logical and sensible than that of evolution and natural selection, and anthropologists must be among the most frustrated people on Earth in not being able to find the definite proof, in bone fossil finds, of human descent from apes. If that were not enough, no natural history museum anywhere in the world can produce the skeletal remains of *any* creature that has escaped the rigid constraints of its genetic code, programmed into its DNA molecule by changing into something else, and this factor is essential to the entire theory.

When the theory became established, many, no doubt, felt a little uncomfortable about being told that they were ascended from apes, but probably felt they would just have to live with it. As a matter of fact, at least one person almost *died* for it when teaching the bold new concept to his class in the USA. He narrowly avoided being lynched by the outraged Tennessee townsfolk and found himself in jail. (More will be said of this gentleman later in the work.)

Apes are gentle, simple creatures and most people would not be offended by descent from them *if* it was proven fact, but currently this is not the case and remarks that could be viewed as a form of anthropological propaganda, such as 'our cousins the chimps' and 'our ancestors the primates', have been made so often that many people are convinced that it *is* all proven fact which no doubt was the intention. Surely, our direct ancestors are Cro-Magnon man. Creative and artistic, he did not interbreed with his predecessor, the squat Neanderthal type, and seemed unrelated to him. If evolution was struggling to produce a human type being, it seemed doomed to failure if the end of the line was Neanderthal man, as he did not seem to be progressing at all. In fact, he appeared to be *retrogressing* and becoming more ape-like and *did, in fact, die out.* Therefore, some would see human ancestry not in terms of millions of years, but in terms of a mere 35–50,000 years.[3] Where did Cro-Magnon man come from? In this regard, could we consider a serious but possible alternative? That Cro-Magnon man was a genetic 'creation' by unearthly beings that stated, 'Let *us* make men in *our* image', a phrase the Bible has (strangely) left in its plural form. It is more like a statement from a board of genetic scientists than a singular statement, such as, 'I will make man in my image' which would be the case if an individual 'divine creator' had made it. I am not the first to suggest this fantastic concept, nor will I be the last and, with many seeing it as pure fantasy, the questions must be, have they looked at the shortcomings of the alternatives? And have they looked at the main and most profound piece of circumstantial evidence, *the awesome human brain?* Some of the questions that I put in this work are phrased in the manner of a suggestion and *may* be viewed as fantasy, hut what is the definition of fantasy? Surely, any theory, proposed but not proven, could be

---

[3] Some anthropologists now put the advent of Cro-Magnon man back as far as 100,000 years.

viewed as fantasy? Therefore, this description fits the theory of evolution of the human being, and as previously said, what could be more fantastic than the Genesis account for human creation?

There is another rather negative and gloomy theory, of course: that all life on Earth is simply an 'accident', perhaps never to be repeated, the assembly by pure chance of all the right type (and the correct number) of amino acids necessary to get things going. But life appears to have such a sheer *insistence* about it, and life's building blocks, in the form of more and more molecular groups being discovered in space, seem to make the blind chance theory most unlikely. And so, in this work, most of the questions asked will be confined to the three main sections asking *did* we evolve from simian creatures, were we '*divinely*' created, or are we children of the stars?

With regard to evolutionary theory as related to humans, there must be many scientists, palaeontologists, biologists and anthropologists who harbour at least a few lingering doubts in some aspects of it. Anthropologists seem to be almost totally united in their acceptance of it, despite the missing and final proof. Why is there fear and reluctance to look at the problems or question its shortcomings by those who remain unconvinced among their ranks?

We might ask, in regard to the theory, is it all over? Who told evolution to stop? Why do we not see living 'stages' of change? Creatures trying to turn into something else? Creation doctrines state that creatures remain 'each unto their kind', and in fact, they do so remain. The chromosome count determining snakes to sea lions ensures that they do. Who were the pattern makers? Who made the moulds? What force determines each species and ensures they remain in that form? When did the process suddenly stop that produced the creature that left the primordial seas and went through all the changes that drawings and animated films suggest, where fins became legs, lizards, crocodiles, dinosaurs and bipedal creatures appeared, and became apes, gradually standing upright as men?

How did all these creatures suddenly become tightly locked into their various 'kinds'? Of course, sketches, cartoons and diagrams and conjecture are all produced to reinforce the theory, but there is a distinct lack of skeletal evidence. Pigs donate their material in cardiovascular operations. No one suggests we came from them. Parrots and mynah birds imitate human speech perfectly, but no species of ape has ever tried. Chimps lack the necessary equipment for speech, even if they had the intellect to try. Having free 'hands' to make tools did not greatly help other creatures with 'hand-like' extremities, especially the apes, who have never attempted to make anything. Yet termites build their air conditioned mounds, bees build their honeycombs, birds builds their nests, beavers build their dams. To be sure, there are many creatures with quite obvious intelligence that would seem to be better qualified to be called our cousins than the chimps. Of course, in a sense, they would be our cousins if they were the creatures selected for the production of the basic anatomy in some great genetic experiment, and would pass on some genetic similarities in *everything except their brain*.

Is it, then, reasonable to ask, 'are human beings children of the stars'? Does our skull house part-extraterrestrial genetic material? Bones of creatures who walked the Earth for nearly 200 million years almost pop out of the ground, enabling us to construct all those impressive skeletons, yet with regard to human evolution, the main question, and certainly the most frustrating one for anthropologists, is why we cannot find numerous complete skeletons of our ancient ancestors, particularly as our museums are so full of ancient animal bones. Only the discovery of more ancient skeletons would finally prove the Darwinian concept of human origins and show a gradually emerging and clearly linked evolutionary line right back to the early hominids.

The only other almost complete skeleton found since 'Lucy' (uncovered in Ethiopia during the seventies) is the so-called 'Turkana Boy' discovered by Richard Leaky in the summer of 1984 on the western shore of Lake Turkana in northern Kenya. The term 'boy' is questionable, because although it was decided he had been nine years old at his death, the remains were classified as being that of a *Homo erectus* from around a million years ago, and the skeleton therefore possessed an ape-like skull. Moreover, the pieces for it were put together from a period of five seasons 'in the field' stretching over more than seven months and involving the removal of fifteen hundred tons of sediment. However, the search continues and as some anthropologists will admit, becomes almost obsessive, so if the links *are* there to be found, such ardent searching should ultimately uncover them.

With the widespread discovery of smaller portions of bones and skulls in other countries, Africa seems to be losing its popularity as the most likely area to produce the final proof. Although not relevant, it is interesting to note that the map of Africa, viewed as the birthplace of humanity, actually resembles the skull of an alleged ancestor, complete with 'brow ridges' at the Horn of Somalia. However, the important point is that human missing links *still* evade the searchers after a century of digging. Are they there to be found?

When we pause for a moment and contemplate the achievements of mankind from the Pyramids to Pioneer voyages around the solar system, analysing closely every planet, walking on the Moon, constructing computers, the Space Shuttle and the Hubble Telescope, planning voyages to Mars and advancing deeply into genetic science, making statements of 'growing' human hearts and organs in laboratories, all brought about by the fantastic human brain, and then look at a chimp's blank stare, a creature who never has and probably never will build anything aft*er 40–50 million years* of evolution, can anyone seriously believe that the human brain is a bequest from *them*?

Ridiculous as it may sound, apes, by now, after such an immense evolutionary period, should have colonised the galaxy. If we have achieved so much in such a short time, what will *we* be capable of (if we still exist) after 50 million years? How can anyone be surprised to see chimps perform a few simple tricks, or pick up a stick to poke an anthill, after such immense evolution, yet we are supposed to be descended from *them*. Behind the likeable docile expression of the chimp lies a brain with only a fraction of the cellular value of human ones, with no intellect, no imagination, no creativity, no sense of destiny and no 'Broca's convolution' in their brain that sets us so far apart from them; no ability (or necessary endowment) for human speech and no hope or wish to be anything more than a chimp.

Underneath the human skull is a fantastic brain with a sense of destiny among the stars, a brain that if not divinely created, could possibly be a direct bequest from unearthly beings that have a huge amount of evolution behind them, living before our solar system was formed.

The human skull contains a fantastically developed brain with a destiny among the stars, with amazing creativity, logic and reasoning powers, *and* the all important capacity for mathematics, which our space activities rely so heavily upon and would not be possible without.

It is estimated that the human brain contains one hundred billion neurons, or brain cells. Interestingly, this is one for every star that astronomers estimate to be contained within our galaxy. It has been calculated that according to the pace of normal evolution, where genetic stability seems to be predominant and positive mutations rare, not enough time has been available since the 'Cambrian Explosion' of life forms for the hugely developed and over-endowed brain to have evolved. There are creatures and fauna, such as fern leaves, frogs, flies, ants, crabs and so forth that look just the same after a time period of up to and over *two hundred million years*. This could suggest that, with regard to the human brain, an exterior force has circumvented a huge amount of time. Alfred R. Wallace, who developed similar theories to Charles Darwin, could not accept that human brain development fitted in with the theory. When his misgivings became public, Charles Darwin wrote and said, 'I hope you haven't murdered our child [the theory] completely'.[4]

It seems preposterous for blind chance to have been responsible for it and, if divine creation cannot be accepted, perhaps all the earthly legends asserting humans were created by gods are correct after all. Genetic similarities in humans and apes would of course exist, if a slowly evolving part-simian creature was singled out for the experiment. It certainly would explain the mystifying appearance of the Cro-Magnon type and the equally mystifying and rapid disappearance of the Neanderthal.

---

[4] Max Flindt and Otto Binder, *Mankind Child of the Stars*, Coronet, 1976.

There is absolutely no comparison between human and simian intellect. One million years ago *Homo erectus* still had the skull of an ape. The current human brain development is only measured in thousands of years. Teaching the apes 'tricks' – to slice through string securing a box containing fruit with a chipped off flake of stone, by striking it with another – is the same as training an elephant to stand on one leg or take a bow. One wouldn't expect to encounter one in the wild that would do such a thing; apes would never *naturally* make stone tools.

All of human achievement may be directed by 'racial memories' and since it has proved that memory can be inherited, are the amazing advancements we are making simply a repeat of things long ago achieved by beings that may have bequeathed part of this supremely advanced organ to us? The very substance we theorise as being used in terraforming processes is found in Earth's most ancient rocks. Was *Earth* terraformed by an intelligence that came this way long ago? Man *will* travel to the stars, find life forms, experiment with them, yet we are comparative newcomers in the galaxy.

Many learned people have theorised that Earth could have been visited by extraterrestrial beings in the past, so we *must* allow for the possibility that this scenario could have occurred right here on Earth. With the many sun-like stars out there, some civilisations will be much older than ours. When hearing Gail Naughton's announcement to the American Academy for the Advancement of Science that our genetic scientists will 'grow' human hearts in laboratories, it beggars the imagination to think what beings a thousand, or perhaps a million years ahead of us might be capable of. Is the alleged UFO phenomena now explainable? Clearly, any unearthly intelligence that may have been responsible for mankind's appearance *would* have an ongoing interest in us. Are we on the threshold of the greatest revelations for 2,000 years, when the descendants of beings that may have been responsible for human creation could be about to reveal our *true* ancestry?

It would cause the most profound cultural shock and social disorientation among mankind ever to be experienced if the ufologists are right after all, and serious information *has* been kept from us. If this is the case, then those that have retained it have done us a great disservice. Such extraterrestrial involvement in human emergence would certainly explain the long and patient observation and guidance seemingly given from the 'angels' of Abraham down to the present day aerial phenomena. Earth, in case, would represent generations of study to curious unearthly beings, with its abundance of life forms, many of which could also have been introduced to Earth, in which case they *would* remain 'each unto their kind' far into the future.

With regard to the apparent insistence of life to proliferate, I recently observed a miniature growth, rather like a tiny tree growing from the top of a large wall. It was a man-made construction of bricks and mortar, yet that tiny bush grew from a seed, dropped from a bird's beak or blown in the wind, that found its little niche and sent out its tiny thread-like roots down into the cavity that no doubt contained the necessary moss and moisture that would sustain it until the food source was diminished and the roots would give up the fight, or find another source. Volcanic islands, hot and barren, arise from the sea and in no time at all are covered with fauna from the same actions of wind-borne seeds and those from the beaks of the birds that alight upon them.

Of course, this only applies to the warm, hospitable womb of the Earth and not the airless, crater-strewn or huge gaseous globes of the rest of the solar system, and when considering life elsewhere in the cosmos, one planet out of nine cuts down the odds considerably. But they will be their all right. With 75 billion sun-like stars out there, they have to be.[5]

Who will represent the seeds, the winds and the birds to bring life to barren worlds in time to come? The answer, of course, will be blue-green algae as the seeds, *our* life form as the birds and winds will be the starships that carry the seeds to terraform other worlds for humans to proliferate upon.

This substance exists in rocks over three billion years old on Earth. Who was responsible for the process on Earth? If mankind is approaching the capability to achieve this kind of action today, then this implies the possible existence of beings over 3 billion years ahead of us. What would *they* now have become.

They would know everything, have achieved everything and probably would not even possess a vulnerable anatomy. There would be nothing else for them to do but create and promote intelligence throughout the universe. *They* would be likened to the concept of 'God'.

Can we believe in an almighty, all-knowing God that created the entire universe in the way specified in biblical teachings? Who put this massively over-endowed intellect in the human skull? When people view the human being as just a refined version of an ape, why would we need such an enormous excess of intellect to survive?

The theories for life's evolution on Earth may be perfectly correct up to the emergence of the pongid apes

---

[5] Isaac Asimov, *Extra Terrestrial Civilisations*, Book Club Associates, 1980.

and onward to the emergence of *Neanderthalensis sapiens*, but the giant leap to Cro-Magnon man may be due to experimentation by aliens, who may also have brought with them to Earth a whole host of seedlings, embryos and plants of all descriptions. As previously stated, this would certainly account for the evidence of so many creatures showing so little change over so many millennia. Back through the mists of time there are various legends of 'star' people introducing various cereals and crops to mankind. If this experiment, we may hypothesise, genetically produced human life forms, their aim would have been to produce the best possible 'creation' in terms of intellect. In this, they may only have achieved partial success with such people as Newton and Einstein. Is it possible we were *all* intended to equal them?

Child prodigies and people with special powers and other exceptionally gifted individuals may also be occasional manifestations of the 'creator's' original intention. Perhaps the positive genes were supposed to eliminate *all* the negative ones, but we are left with this 'Jekyll and Hyde' condition in our mental processes, with the dark side holding us back but fortunately not gaining a dominant situation in the brain.

Consequently, just as we step outside the normal endowments of the brain, with the occasional geniuses on the positive side, we also have had the evil and negatively-minded moving among us, such as Adolph Hitler and many that preceded him. if extraterrestrial beings *are* overseeing our development and behaviour patterns resulting from the actions of their forebears, they may be anxious to instigate some form of 'genetic correction plan' to bring us back to the original intended specification, perhaps by some form of neurological 'adjustment'.

I said in a work called *Cultural Shock* that some people, after a fall, have developed strange powers resulting from a blow to the head. Surely, this implies that the necessary neurons were 'bridged' by the impact, and if we knew which circuits to bridge, we could *ourselves* produce all kinds of geniuses. Similarly, in cases of insanity, we could close off a few circuits and thereby perhaps be able to cure such unfortunate people. These operations may be seen as simple neurological adjustments by beings many centuries in advance of us.

There may be beings that could produce any life form they chose, given the advancements we ourselves have made in such a comparatively short time They may also be able to manufacture or create anything they wished, or rearrange matter by changing its 'atomic number'. We actually envisage doing this ourselves, as we look deeper and deeper with ever-improving microscopic enhancing equipment into the heart of the particles making up all matter. Silicone chips are becoming so miniature they are practically disappearing and the gaps grow larger and larger between peoples still living in the Stone Age in such places as the Australian Bush, New Guinea and the Amazonian interior, with other nations planning trips to Mars. But, of course, the same situation existed 4,000 years ago with the comparison between ancient Egyptians and ancient Britons *and* a lot of other nations of the world at that time.

The human brain *is* a supremely over-endowed cellular arrangement, and in people like Newton, Rutherford and Einstein, it could only be described as *massively* over-endowed. This is what caused Albert R. Wallace to waiver over the human evolution theory, for the simple reason that it defies the natural selection process which is only supposed to endow a species with sufficient intellect to survive.

The human being has intellect far in excess of its simple need to survive. Even if none of the aforementioned had happened and eventually all the necessary bone fossils and halfway stages are found, mankind's destiny will still be among the stars. We may be the 'first'. It may be entirely down to mankind to 'go forth and multiply', and by terraforming processes bring life to, *and* populate other worlds. if the unidentified aerial phenomena *are* misunderstood natural phenomena, we may very soon have a real phenomenon to deal with as our radio waves are now sweeping over possible worlds to a distance of 70 light years into space. Intelligent beings may even now be laying plans to visit this obvious source of intelligence. Have *they* been very clever in blanketing their *outgoing* radiation while developing their receiving equipment highly? Now 'they' all know we are here, but we (after 30 years of searching) have not hear a peep out of anyone else.

However, there is so much circumstantial evidence for a long, continued presence of intelligently controlled objects in our skies, we have to consider whether there will be a culmination of it and when. Will it be the 'Great king coming from the skies' who Nostrodamus predicted for 1999? Have they a long term four-phase plan in operation:

1. Initial creation;
2. The emissaries to teach and guide us in Abraham, Moses and Jesus;
3. Abductions to see where they went wrong in witnessing our crime and wars;
4. The final second coming?

Section I
Human Evolution – Questions

1. Disappearing Entities

Why did the hominids disappear? They were supposed to be the first link in the chain of the advent of creatures with some human-like qualities. If this is so, then they should have been a more advanced life form than the apes. Apes are still with us. The hominids have disappeared. They reigned for at least 12 million years and in all that time apparently did not develop any attributes to establish their superiority over the true pongid apes. They existed for a time period of some 12–14 million years, then came the emergence of *Homo erectus* around a million years ago. Therefore, in a comparatively short time of 600,000 to one million years, humans have developed to the point of contemplating interstellar voyages and have scrutinised at close quarters our entire solar system, but in *their* enormous evolutionary period the hominids achieved nothing and hardly developed at all.

Homo erectus was supposedly a major step forward with his erect posture and, though apparently having an ape-like skull, was allegedly the first true human-like being. Why did *Homo erectus* fail to flourish instead of lasting for a 'mere' 500,000 years? Surely *he* was better equipped than the hominids. There is no obvious reason for his demise –

another supposedly improved human-like version disappearing. Then comes Neanderthal man. More human qualities, a larger brain still, yet only existing for the extremely short period of 75,000 years – a blink of an eye in cosmic terms. Finally, our true ancestors, Cro-Magnon men, appear. Various beings appear and disappear within this procession, all seemingly unrelated to each other. Could natural selection and evolutionary processes alone be so rapid and selective from *Homo erectus* to modern day humans, when it still had dinosaurs languidly chewing foliage after 180 million years of totally purposeless existence?

What other force of a more determined nature could have been responsible for the disappearance of these beings?

Actually, although being superseded by *Neanderthalensis sapiens* some 500,000 years after his appearance, *Homo erectus*'s actual *disappearance* seems to have occurred after only 300,000 years and produces a questionable gap of 200,000 years before the advent of Neanderthal man. Clearly, the most massive change in bone structure, posture, cranial formation and shape all took place a mere 30–50,000 years ago, and give the strong impression of a creative exterior force influencing such rapid change.

Regarding creatures that have remained the same for so long, there is a species of octopus found in fossil form, looking just like a modern day octopus, but dated as *125 million years old*. Profound events in selective and rapid change have only occurred in the transition period between Neanderthal man and Cro-Magnon man, with amazing changes in brain development. The Neanderthal brain was actually slightly larger than modern man's, but as mentioned in the work, this may be as irrelevant as comparing the brain size of an elephant to an organised constructive intelligent ant. In *The Book of Life*, edited by Stephen Jay Gould, it is made clear that the Cro-Magnon people were contemporaries of the Neanderthals for a time and were *not* their descendants, and asks where they came from.[1] An interesting question, and precisely what this work is about. It also points out that there was no evidence of conflict between the two beings making the disappearance of the Neanderthals even more mysterious.

2. The New Messiah

When Charles Darwin first introduced his theories to the world, it was a brave and daring act for a man of his time. It was a rather staid and somewhat stuffy era, with established ways of thinking and, of course, the church's power and influence was far stronger in those days. Small wonder that the theories outraged the religious factions.

Even the esteemed American lawyer, Clarence Darrow, could not completely win his case in the famous 'monkey trial' of 1925, when defending a certain Mr John Scopes who had taught the theory in his school and quickly found himself in jail.

A certain gentleman appeared who was quite affronted by Mr Scope's teachings in the shape of a Mr William Bryan, with a fire and brimstone manner, who managed to whip the Tennessee townsfolk into a religious furore. Mr Scopes was almost lynched for teaching a theory, which seemed to him to be quite logical, although at odds with established religious thinking. He was sold lock, stock and barrel on the 'theory'. I emphasise theory, because this word is quite often dropped from he phrase, 'the theory of

---

[1] Stephen Jay Gould, *The Book of Life,* Ebury Hutchison.

evolution and natural selection', and we just hear references to evolution. If ever a theory deserved to be true it is this one, as it seems so perfectly logical and almost certainly is, at least with regard to creatures of the world up to the true pongid apes.

But, of course, there are problems. We cannot find the bone fossils to prove it all, especially with regard to human development, and as bones clearly survive for immense periods, as is evident in our dinosaur collections, we should be able to find halfway stages of one creature changing into something else. It has to be said that it must all, for the moment, remain a theory. Certain states in America today (a kind of victory for Mr Bryan) do have their Boards of Education specifying that new textbooks should also contain the other theories of human origins. This would obviously include divine creation, but whether (in spite of all the convincing literature on the topic) extraterrestrial theories regarding the intervention in human development are also included there, remains doubtful. However, the Darwinian concept has in some areas apparently lost its 'unassailable' image, and, as said, some anthropologists are honest enough to admit that the entire theory is not clearly proven with regard to a clear evolutionary process from the primates.

The aforementioned Mr Alfred R. Wallace, who was the co-developer of the theory (and is generally little mentioned with most of the credit going to Charles Darwin) was plagued with all this uncertainty. Later on Mr Wallace became critical of the theory, largely because of the vast difference in the development and power of the human brain in comparison to apes. He felt that spiritual forces were the only thing that could account for the development of such an organ with its massive 'over-endowment' of intelligence.

The evolutionary theory should be fact. It seems so logical, particularly as it seems all creatures share the same common gene, and it could be true regarding other creatures. Man is supposed to have developed from a kind of lung fish that decided it would leave the sea for life on dry land. It is easy to imagine in the mind, particularly as we have had help from pictorial diagrams and moving pictures of a fish forming lower fins into stubby legs, poking its nose, out of the water, then finding its locomotion ability, then stretching into a lizard, or a crocodile shape, then gradually raising itself on to hind legs, becoming a dinosaur type eventually becoming more erect and forming into an ape and then a man. It is these types of drawings and cartoons that are largely responsible for fixing the idea in people's minds that it did happen that way, but films and diagrams are one thing. Hard facts, in the form of fossil links, clearly showing one creature changing into another, i.e. halfway stages, are quite another.

For well over a hundred years, the patient, dedicated and hard work of numerous archaeologists, palaeontologists, anthropologists and biologists who have sifted dug, examined, theorised, dated and drawn, logged, catalogued, has given us so much information that books on it abound and fill the library shelves. Yet, alas, the end result still remains inconclusive, a fact which renowned anthropologists (such as Richard Leakey) freely admit.

We know that the true pongid apes (as opposed to hominid types that are alleged to have some human traits) wandered the Earth 30–50 million years ago. But the first alleged hominid type, 'Ramapithecus',[2] who existed some 14 million years ago, does not have a clear fossil lineage down to modern man by any stretch of the imagination. In fact, there is an enormous period of time (around 12 million years, almost the entire Pliocene period) with hardly any fossil fragments to illuminate the theory.

It is almost certain that if Ramapithecus existed today, he would be in a zoo. True man, in his development, can only be allowed a maximum of about one million years since the appearance of *Homo erectus*, or only a minimum of 110,000 years since the appearance of Neanderthal man. Some would go further and state that 'true man' was really Cro-Magnon man appearing a mere 35–50,000 years ago. From the Australopithecus genus down to the appearance of modern man, including *Homo erectus* and *Homo habilis* (or man the tool maker) there is no clear fossil lineage, but certain types appear with no distinct links between them, and we struggle to find further proof in bone fossil evidence that shows a direct connection between the hominids and modern man.

Some theories of man's early beginnings in Africa with regard. to hominid types are not based on entire skeletons or even complete bones, but *fragments* of bones.

The enormous gap between Ramapithecus down to modern man seriously weakens the theory, as there seems to be no common ancestry between apes and men. The theory has been in a state of flux since its inception. There has even been a complete invention, innovated and given the name 'proconsul' and theories proposed on this hypothetical creature without the necessary fossil links in existence. Of course, the entire evolutionary theory with

---

[2] Identification as hominid now doubted by most anthropologists.

regard to the emergence of mankind was not due to fossil finds encouraging its formation, but the other way around. The theory was set down on the assumption that the links to prove it *would* be found. Indeed it seems such a logical theory that they should have been found. Of course, all this is only in regard to human skulls and the skeletal frame.

But, it is what is *inside* the human skull that utterly confounds the theory when we move into the natural selection part of the theory. There is so much diversity of opinion with regard to finds and what they are supposed to prove. Some even theorise that the links, so vital to the theory, could be anything up to 50 million years old. It is rare to find experts in the field who entirely agree with each other. One wonders what Darwin's opinion would be today if he were alive to comment. There is only one obvious human body shape emerging with the advent of Cro-Magnon man. There is an enormous variety of apes and monkeys. Can we be really sure that hominids and, in fact, all the beings before Cro-Magnon man were not just another type of a more refined ape-like creature? The so-called missing links, if ever found, would have to show skulls clearly changing to human types (which differ greatly from ape types *and* so-called hominid types.) And, most importantly, there should have to be conclusive finds showing hip joints changing to give modern man his upright stance.

The drawings, so often depicted in animated films and so forth are merely conjecture but many people in a kind of 'brainwashing' process assume it actually happened, not quite realising that the theory is still not entirely proven fact. To discover the compelling fossil evidence remains the dream of every anthropologist. Furthermore, other creatures seem to remain much the same in their general appearance, going back millions of years into the past. Another mystery is the so called 'Cambrian explosion', where fully formed and developed creatures seemed to appear all together, with only low life forms preceding them and fully formed marine creatures with *no* preceding fossil history. In this regard, 'divine creation' seems to have more going for it than evolution, with so many creatures remaining 'each unto their kind'. Although pre-Cambrian jellyfish types existed, species with skeletal frames seemed to 'burst' into existence.

The time period given for the development of mankind from a savage grunting sub-species to walking on our moon seems a ridiculously short time for the emergence of such a reasoning, thinking, intelligent being. Even more so, with regard to the rapid evolution of the brain, so over-endowed and developed in such a short time span. The whole point and lynchpin of the theory centred around Darwin's assumption that the fossil links *would* be found showing one creature changing into another, and there are no such links or bone fossils existing in the record.

3. The Hundred Year Quest

According to *The Story of Archaeology* (Wiedenfeld & Nicholson, 1996), Louis and Mary Leakey 'constructed' a skull from 400 pieces of bone. It was also stated that thousands of pieces of animal bones had been found at the site in the Olduvai Gorge in South Africa in 1959. Can we be assured that *every single one* of the fragments used were subjected to accurate identification and dating methods just coming into being in the fifties, such as the Potassium argon process?

Another fossil, 'No. 1470' in the Kenyan National Museum, was found a few years later and was stated as being 3 million years old, then revised to 1.9 million years later on (a difference of over a million years).

The 1959 finds were accidentally discovered when Mary Leakey was walking her dogs, then were reburied until a film crew arrived. What's more, a discussion between her and friends revealed a previous intention to make a film, and so the film crew were summoned to record the rather coincidental 'find', exactly one hundred years after Darwin's publication of *On the Origins of Species*.

There seemed to be a determination to bear out Darwin's assumption that Africa would produce his required 'links'. However the skull, when constructed, was very gorilla-looking with a ridge on top and, of course, the usual huge eye sockets. Only the teeth seemed to convince them of some human-like characteristic. When we consider the 192 different species of primate compared to the one single identifiable human skeleton of the Cro-Magnon man being, our only convincing ancestor, can we be assured that these grotesque skulls were not simply another variety of ape?

Scholars can be wrong. It was 'scholarly opinion' that rejected the so-called 'Taung Child' found in 1925 on the fringes of the Kalahari Desert, (probably quite rightly) as it seemed to be a young chimpanzee, but the 'scholarly opinion' did their rejecting on the basis of the 'Taung child' not resembling the half-human/half-ape characteristics of an earlier 1912 'find' in Piltdown Sussex, *which later proved to be a hoax.*

'Scholarly opinion', more recently, acclaimed the alleged 'Hitler Diaries' as genuine which also proved ultimately to be a hoax. The respected Arthur Conan Doyle believed, wholeheartedly, in the 'Cottingley Fairies' until one of the elderly perpetrators confessed to photographing paper cut-outs with a Box Brownie before her demise.

Scholars are not infallible beings. Surely, when the driving motivating force is upon them to find the ancient human ancestor by constructing the hoped-for links from tiny pieces of bone, where thousands of other animals' bones are evident, including no doubt those of apes, there is an amazing potential for error.

'*Homo habilis*' is still widely disputed regarding his fitness to belong to the human genus. Although this work looks at some of the potential for error and highlights the questionable use of the phrases, 'our ancestors the primates', and 'our cousins the chimps', it is still, in spite of the lack of convincing evidence, not refuting the hard work of learned people and possible evolution from an ape-like ancestor. It only draws attention to the strange, 'brick wall' that human evolution comes up against after the millions of years of assumed evolutionary effort, of various beings appearing and disappearing with little connection between them, only to stop *dead* at *Neanderthalensis sapiens*, having no connection with the superseding being, i.e. our immediate ancestor, Cro-Magnon man.

Certain works state that the apes 'make' tools for immediate use such as stripping leaves of twigs to poke anthills, but they no more 'make' the tools than certain amphibians 'make' the stones, they *utilise* when lying on their backs in water and placing them on their tummies to crack open shellfish.

The march towards the doomed *Neanderthalensis sapiens* really began with the discovery of *Homo erectus*, appearing around 600,000 to one million plus years ago. Apart from the 3 million year-old footprints and the allegedly upright walking 'Lucy', *Erectus* seemed to be the indisputable proof of an erect semi-human being, yet he had a small brain and a skull like an ape. Various other puzzling finds, skulls and bits and pieces, have been found and argued over by the anthropologists, but the sum total of all the finds remains tantalisingly small.

However, in spite of the lack of convincing and final proof, human evolution does seem to have been striving to produce the seemingly ideal body shape and appeared to be crying out almost for a 'helping hand', particularly as *Erectus* had such a small ape-like brain and all the millions of years prior to *his* appearance had not helped at all in the development high intellect.

Currently, most anthropologists would probably agree that the human fossil collection is frustratingly sparse. We do not have *one* complete skeleton; instead we have 40 per cent of the skeletal frame of small female creature christened 'Lucy' and alleged to be an upright walker that was found in Ethiopia in the Seventies, although it had the teeth and skull of a chimp, long arms and curved toes, suggesting a tree living habitat. It was dated as some three million years old. (The only other incomplete skeleton is the 'Turkana Boy' mentioned in the introduction.)

A similar date was assigned to footprints found in Tanzania, suggesting 'upright walkers' (leading up to the doomed *Neanderthalensis Sapiens*) having existed a lot longer than previously thought. They caused great excitement among anthropologists, but with regard to the amazing bootprint found in sandstone by a joint Soviet/Chinese palaeontological expedition and a *shoe sole imprint* showing faint traces of stitching from the *Triassic*[3] period in Fisher Canyon, Nevada, they are strangely silent.

---

[3] Around 200 million years ago.

Apart from the hominid and other skull portions, the rest of the collection are mostly scraps. Great efforts are being made to vindicate Charles Darwin's assumption that Africa would be the birthplace of humanity, indicated in his 'Descent of Man', published in 1871.

The hominid skull parts, such as *Australopithecus africanus*, found in South Africa in 1947, and *Australopithecine paranthropus boisei* (or the 'nutcracker man') found in the Olduvai Gorge in Africa in 1959, are extremely ugly and unhuman-like beings, and how anyone could assign human-like qualities to the so-called 'Black Skull' discovered in sediment in Kenya is beyond belief. It has a broad, flat face with high cheekbones, huge raised eye sockets and an enormous ridge on top of the skull. On top of this there are no fossil remains at all showing a link between the true pongid apes and the hominids which by their appearance seem to be just another species of ape.

At the end of the line, we find 'Neanderthal man' contemporaneous with and *not* preceding Cro-Magnon man. The anthropologists are struggling to fill the enormous fossil gap from the earliest hominids of 14 million years ago. The so-called 'Taung Baby' found in 1924 is considered to be too young and undeveloped for a proper assessment of its *final* appearance. In any case, its face and brain are 'ape-like' and so, when we look at our rather sorry collection of bone fossils today, gathered in the search for the elusive links to definitely prove the theory of human evolution, they hardly seem to represent over a century of hard and dedicated work in sometimes hostile climates, and even the ones we have, in some cases, seem more plaster to neatly fill in the missing bits than actual bone.

It is interesting to reflect that today, many scientists have trouble maintaining their pet theories and finding support for them and even suffer derision over them, yet Charles Darwin was able to formulate a theory without any evidence (rather like a chief of police grabbing a person off the street, pronouncing him guilty of a crime and telling his men to go and find the evidence), and consequently a veritable army of anthropologists, biologists, anatomists, archaeologists, students and labourers would sally forth and patiently dig, sift and scratch about, categorise, formulate and scientifically date for over a century to try and prove his theory. He must have been quite a charismatic personality. However, he *did* provide an acceptable alternative for many, who promptly rejected the mind-stretching pronouncements in Genesis.

As previously mentioned, the words 'the theory of' have been rather subtly dropped from the sentence, and today we just hear 'evolution of the species', as though it was all a proven fact, and this is certainly not the case.

Most people would have no problem at all in living with the fact that we did evolve from apes, as long as it was finally proven and all the conjecture was removed, but it has to be said that no definite proof has been found regarding human evolution or fossil links showing other species changing into something else, through long time periods.

Apes are fairly gentle creatures, with social and family organisation, and, by and large, behave better in some cases than people. (Some would go further and say that human beings, in some cases, appear to be the most barbaric creature ever to walk the Earth. When we are not raping and pillaging each other, we unite to perpetrate the same crimes against the Earth and our environment.) To be sure, if definitive proof of a conclusive 'halfway' stage between man and ape *was* found, most people could quite happily accept it, as long as it *was* a substantial fossil find and not something built up out of conjecture from a tiny piece of bone.

Ardent supporters of the theory revere Charles Darwin, and to say he might be wrong would be quite unthinkable to them, and so the quest goes on and on. The masses of bones quite easily found of creatures that perished 65 million years ago *must* seriously embarrass the searchers. Many people avidly watch the natural history programmes produced quite professionally on the television, but nothing startlingly new ever seems to crop up with regard to fossil finds, just the same familiar theorising and conjecture on the latest small fragments.

One may recall just such a programme which showed the eminent son of an equally eminent father puzzling over a large box of tiny pieces of bone with his wife and a friend. He turns to the friend and says something like, 'I cannot find two pieces that even fit together. In fact, I'm fed up to the teeth with it. Here you are,' (shoves them over to his wife), 'you have a go.' Later of course, lo and behold, there is a skull, neatly built and resembling something 'halfway' between a man and an ape. Suddenly all the pieces that had so puzzled the 'expert' now fitted neatly together and resembled the sort of thing they were hoping it would turn out to be.

Now one is not implying that the good lady filed a bit here, gouged a bit there and got rid of an 'irritative' point or two, or made the odd recess. I would simply assume that, unlike her husband the 'expert', she was just very good at jigsaw puzzles.

Many find it singularly irritating when they hear on such programmes, 'our ancestors the apes', and 'our cousins the chimps', when there is no clear and decisive evidence to prove the 'theory'.

It is a long time since the Piltdown Man hoax of 1912, when the desperate urge to prove the theory tempted people into artificially ageing and burying bones that almost fooled those who wanted to believe. I do not doubt the integrity of today's teams of palaeontologists, anatomists and anthropologists, but Murphy's Law states simply, 'if a thing *can* happen then somewhere, somehow it *will*', and we must remember that, if they were desperate *then* to prove the theory, how desperate are they now 83 years later with much cleverer techniques at their disposal.

We have a hard time today to know for sure if antiques, watches, or even designer clothing are the 'genuine articles', so to speak. When we see the range of skulls with ridges on top and huge eye sockets that are supposed to have human characteristics, and note how much plaster is included in them to complete the build, the only conclusion can be that they were apes and nothing else. To state that apes have a large percentage of the genetic material that humans have, does not really support the theory because all creatures have a certain percentage of similar genetic material, which may even include life forms on *other worlds* who will be made up of atoms and *stellar material*, just as we are.

Considering the more remote lineage of apes, it is surely strange that they are *still* apes, with no sign of them changing into anything else. Because of this long existence they must have cohabited with mankind and not viewed as enemies. Therefore, their bones would be disappearing into the ground along with human remains and many years later these bones, both human and ape, would have been found jumbled up together and quite likely would have had similar ages when subjected to our dating processes. They must have run the risk of winding up in the same box of bits and possibly even have been stuck together to form a skull by some less competent or over-zealous researcher.

Written works on the evolution of the species are riddled with conjectural phrases such as this or that 'must' have happened, and that 'must' have changed, but the biggest stumbling block, that of the genetic code, or 'chromosome count' in a species preventing change, cannot be overcome lightly. All creatures of Earth would have common genes. What are genes? Material made up of atoms, 'cooked up' in a star long exploded. All creatures that may exists on all other worlds would be made up of atoms and no doubt have *genes*. Therefore quoting percentages of genetic material seems meaningless. We might as well say that, as parrots can talk, this is a sure indication that we came from them. Before one laughs too loudly at this, a gentlemen by the name of Professor Huimar von Ditfurth, in a book entitled *Children of the Universe*, stated on page 190 that mankind and the *chicken* had a common ancestor *only* 280 million years ago.[4]

Half human, half ape life forms could never be produced in a *natural* manner, and if it was attempted we certainly could never call it natural. But if the abhorrence of such an act could be overcome, a human male could couple with a female chimpanzee, our so-called *cousin*, from now until doomsday and no quasi-human offspring would result, simply because the aforementioned code ensures protection from the chaos that would result from inter-species copulation.

---

[4] Professor H. von Ditfurth, *Children of the Universe,* Futura, 1975.

A popular phrase used in evolutionary parlance is *branching off* – that the human form *branched of* from the primates long ago. This is rather a glib term for such a monumental event. Surely *mutated* would be a more fitting phrase and, since mutations are usually brought about by forces *disturbing* nature such as excessive radiation (which is theorised to have caused thicker dinosaur egg shells, preventing their hatching out), and are usually harmful to a species and not advantageous. If so, then this *branching off*, in theory, would not have produced a better equipped or more advanced species, but a lesser variety. And if this *branching off* occurred to produce the hominids, *Homo erectus* and Neanderthal 'man', it clearly failed, as all these beings *died out*.

One would imagine that if fossil links ever existed showing change or some intermediate stages of the process of one species turning into something else, they would by now, after all these years digging, have been found. After all, as previously said, bone material lasts for a very long time in the ground.

The aforementioned alleged human-type entities that superseded their predecessors seemed to disappear more quickly with each successive being. Hominids lasted 14 million years, *Homo erectus* 500,000 years and Neanderthal man 75,000 years. With this massively reducing trend, the Cro-Magnon man from which current humanity stems must be due to disappear *any time now*, and since no other being of a more refined variety exists to take our place, *could this be the end of the human race?*

Even if we take evolution right back to the beginnings of the solar system, we are basing everything on assumptions. We assume that the same material that formed the sun broke off in swirls to form the planets. Now, whereas this is believable in the big gaseous giants, what about the rocky worlds? Where did all the heavy elements come from? There are only minimal heavy elements such as metals in the sun, which is composed mostly of hydrogen gradually becoming helium. Logically, one would expect *all* the planets to be gaseous, like failed suns, and similar in make-up to Jupiter, so it can be seen that questions and conjecture exist right back to the creation of our star system, let alone the evolution of our species.

If we return to the box of bones our famous anthropologist puzzled over, there must have been a couple of hundred small pieces. Now, the carbon dating system is only good for up to 50,000 years,[5] and utilising other dating processes must be a costly business, so one finds it hard to believe that *every single* one of those small pieces in the box (about two foot by one foot and about four inches deep) was subjected to an elaborate ageing process. Therefore, although scratched out of the ground in the area of the original find, some may differ immensely in origin from other pieces if that area or habitat was favourable and occupied for a long period of time, and surely other creatures' bones and the remains of chippings and the broken bits and leftovers of tool manufacture, such as bone cups, bowls and other items, would permeate the site, not to mention bone material in their food waste.

Those dedicated to the Darwinian concept would refute 'divine creation' as readily as they would extraterrestrial genetic creation, but believers in 'divine creation' would point out that mankind did appear 'suddenly' upon the Earth and that fossil finds only serve to confirm this fact.

---

[5] This is an important point, and indeed it has been stated that none of our dating processes, including the 'TL' or thermoluminescent dating process are 100 per cent foolproof and reliable.

Those people advocating extraterrestrial involvement in mankind would certainly agree with this 'sudden' appearance, but would point to the unrelated types appearing, with circumstantial evidence in this case seeming to indicate some form of 'experimentation', and in particular the rapidly evolved and unnecessarily over-endowed brain, in no way conforming to slow, plodding evolution. This brain blossomed into existence in *thousands* of years when it should have taken *millions*.

Other facts that could be seen as circumstantial evidence are the many earthly legends of human beings having resulted from 'sky people' and 'gods' *creating* them. Of 'initiators' arriving to impart wisdom and knowledge of farming, town planning, hygiene and good government. Also numerous references, steles, cave drawings of beings in spacesuits and humans in kneeling positions looking up in awe at orbs and discs, as though communicating with unearthly beings. It is perfectly possible that the human brain, enabling us to achieve our level of technology so quickly, *is* a directly example of God's handiwork. But He would also have to be responsible for our darker, negative traits that cause all our crime and continuing wars and are still active within the brain which many would prefer to attribute to more *fallible* beings, and extraterrestrials, in spite of centuries of advancement and creativity (all passed on to us), would still surely be *fallible* beings.

Clearly, when observing other creatures, quite obvious signs of intelligence and creativity become apparent in the aforementioned termites, ants, bees, birds and beavers, who quite ably create things that are necessary to their survival. Also it is clear that such organisation and creation implies good communication abilities. However, for all the obvious intelligence displayed in other species, they all function within the parameters of natural selection theories, within their restricted environment, and have only been endowed with sufficient and no *excess* of intelligence to contemplate on or to achieve other things outside that certain environment. One could not, for example, imagine a termite colony conspiring to mine out and forge metal deposits and start building a spacecraft. Why has mankind's cellular organ inside his skull been so massively and obviously bestowed with intelligence far in excess of his need to survive in a primitive environment? More will be said of this massive over-endowment, but we are taught by evolutionary concepts that it is something the processes of natural selection are *not supposed to do*. Normal evolution and natural selection should simply have ensured that human beings advanced to the point of becoming planters, hunters and gatherers rather than striving to reach (or return) to the stars.

When we look at the minuscule collection of bone fossils available to study after all that hard and dedicated work of diligent digging and sifting, it certainly does not seem, as said, to represent a century of work in the field, and one wonders if any positive links will *ever* be found. The only changes in a species we seem able to point to are within kind, that is, adaptations or improvements helpful to that species. It would be really quite amazing that if the only thing Mr Darwin had to go on that encouraged him to surmise we evolved from them was the faint similarity to humans of an ape when standing upright (for the short period its hip joints will allow), but before the business of digging for the evidence really began we must assume that this was the case, as a good deal of the meagre fossil evidence of ancient humans was found after the theories were formulated. Geological evidence seems to suggest that violent holocausts have occurred in the past, periodically wiping out almost all living creatures around at that particular time period, so perhaps if we dug in the most unlikely places we may get better results but then – of course, such geological upheavals would have buried the many, quite easily found, dinosaur bones. Therefore, it would appear that nothing *too* geologically catastrophic has occurred in the last 65 million years since *their* extinction.

All the drawings and the manufactured models in our 'times past' exhibition of ape-like brutish creatures camped around a fire wearing skins and grunting to each other, may for all we know show what *they* were 'reduced to', rather than what we 'evolved from', for there is nothing more certain than the possibility that we, in spite off our modern day technology, could be quickly reduced to that state again in the aftermath of a global nuclear holocaust. We would kill dogs for their meat and fight over their bones, and all our thin veneer of civilisation would disappear at a stroke. So it can be seen, that in spite of the amazing human intellect, a precarious balance exists between civilisation and savagery.

To contemplate the very rapid advancement of modern man in the evolution of our species from the hominids, it seems fantastic that such 'shape-shifting' in bone movement could occur in the skull almost immediately to produce Cro-Magnon man. We could look at a fossilised frog from the Cretaceous period of 170 million years ago, and it is *still* precisely a frog. Ants and spiders trapped in amber, look exactly as they do today, with jellyfish, the same after 600 million years, fish after 500 million years, fern leaves after 280 million years and so on.

If we are happy to accept that we came from apes, we must decide what type we would prefer to have evolved from. In humans, there are differences in colour, build, height and so forth, but the skeletons are still the same, but there are 192 living species of primates, apes and monkeys.

When we consider all the simian species from spider monkeys to gorillas, chimps, orang-utans and so forth, all happily existing within these types and showing no sign of changing into anything else, the theory seems more than a little shaky and most certainly encourages the hypothesis of extraterrestrial 'involvement' bringing about rapid change with each genetic 'experiment'.

The fossil record as it stands seems to match this hypothesised 'experimentation' more neatly than a natural evolution of one primitive type moving towards the more advanced type in steady procession over the enormous time periods that natural evolution seems to demand.

If an ardent supporter of the human evolution theory was asked if he or she believed other-worldly intelligence existed, they would probably say *yes.* If asked if he thought they had ever visited Earth, they would probably say *no* (not willing to be drawn into that trap). If asked if *we* will ever carry out advanced genetic experiments, they will probably say *yes*. If asked whether it was possible other-worldly beings had done so *here*, they would probably say *no*. If asked would our future astronauts ever help a species to save itself from extinction by simple genetic experimentation or guidance on a distant world, they would probably say *yes*, and so one can see emerging an obvious contradiction in consistency, controlled by a central belief 'blinding' one's open-mindedness to other possibilities.

When mentioning the various species of monkey bones that have gone into the ground, we must not forget the gibbons, the red mantled tamarind, the Bornean Proboscis monkey, the blue monkey, the red howler monkey, Verreax

Sufaka and associated primates, such as the Maholi bush baby, the ring-tailed lemur, the Philippine tarsier and so forth. Of course, all these jigsaw puzzles with bits of bone, and the building of human skull, are nothing compared to what is housed *inside* them. In all the time since the early emergence of life forms hundreds of millions of years ago, the brain of the ape can only muster one billion neurons with no sign at all of a 'Broca's convolution', a sure sign of intellect in a human brain with its 100 billion neurons as compared to the poorly endowed ape. In actual brain weight the human brain weighs in at 1,375 grams to that of a chimp's 400 grams.

Species that have intelligence, as said, certainly communicate but speech will always elude our friends the apes without this essential convolution. Not only that, they also lack the necessary vocal equipment in their throats. Knowing these facts (as one assumed they did), it is hard to understand why the experimenters wasted their time taking chimps from birth, putting nappies on them bottle feeding them, talking to them every day and expecting them to turn into a sort of quasi-human child and start asking questions. The whole experiment was doomed from the start.

If human beings *are* an 'experiment', then, even allowing for this massive intelligence in comparison to simian types, we could hardly be classified as an unbridled success. Ever since the earliest records of human history and advancement, we have been slaughtering each other and continue to do to this day so obviously, along with our positive attributes, we have an equal amount of negative and destructive traits locked together in a 'good versus evil' conflict. Perhaps further activity in the 'experiment' is required, and long overdue.

It is clear then that mankind is definitely an anomaly, a misfit in the evolutionary plan. No other species on Earth, as said, has been so mentally 'over-endowed' far in excess of its need for survival, particularly with regard to the reasoning powers, self-reflection and powers of abstract thought processes to contemplate not just our immediate surroundings, but the infinite universe. By no stretch of the imagination could one ever envisage an ape sitting down and thinking anything out or even contemplating his next meal, let alone wondering about eternity, for instance, or what is on the far side of the moon.

The gulf between humans and the apes is enormous and insurmountable, but this massive intelligence should alarm us rather than make us proud of ourselves and, whereas genetic experimentation is always taboo unless connected to safe areas such as the prevention of Down's Syndrome births or other mistakes of nature, it should, and probably will, in the near future be extensively used for the elimination of savage and aggressive genes which are useless to us if we are ever to develop into *truly* civilised beings. Even now geneticists are slowly beginning to isolate the genes for various human traits, so the process has already begun.

To return to the subject of the Darwinian theories and modern fossil finds, there are still more questions to be asked regarding the evolution of mankind. After the disappearance of *Homo erectus* 300,000 years ago (for no obvious reason), there is a questionable fossil gap of nearly 200,000 years before the appearance of *Homo Neanderthalensis sapiens*, appearing as he did 110,000 years ago. Again, there is a series of rather inexplicable jumps forward. However, the word 'forward' may not be a suitable one due to the more 'modern' find of the 250,000 year-old Swanscombe skull portion, allegedly similar in appearance to modern man's. What is it doing in that time period? Almost like reverse evolution, a more human-like creature was superseded by a less advanced being. The general view of Neanderthal man is that he was squat, muscular and beetle-browed, yet his brain case was larger than modern man's by about 100cc. However, analysis of the inner cranial structure does not suggest he had the capability of speech. Neanderthal man also disappeared from the scene about 35,000 years ago, with the advent of the obviously more intelligent Cro-Magnon man, but still reigned for 75,000 years before then. The ice ages of the period cannot be blamed for his demise completely, as his type occupied the Mediterranean and other areas unaffected by the ice sheets.

Supporters of the alien genetic experiments theory have much to seize upon with these many and varied *Homo* types suddenly appearing and disappearing up to Cro-Magnon man. Of course, the appearance of Cro-Magnon man is something of a mystery in itself, due to the absence of fossil finds linking him to Neanderthal man. When one considers the Cro-Magnon culture, in particular his magnificent cave art, there is no doubt he would have found the comparison between himself and Neanderthal man just as insulting, as some people feel today when being told they are related to chimps.

Cro-Magnon man also had a brain case larger than modern man's but brain size, even within species, may be irrelevant with regard to the intelligence therein, as when comparing a human brain's size to that of an elephant. There certainly seems to be a good case for the 'induced retrogression' concept regarding Neanderthal man, as the later finds him appearing more brutish and caveman-like than earlier ones, and show him becoming more primitive and not evolving at all. *Was this purposefully introduced retrogression?*

Darwin explains natural selection as a force that is continually scrutinising for failure, rejecting it and enhancing the good qualities, yet anthropology generally regards it as an impersonal and quite random process. But it seems Darwin was implying that it had a purposeful and intelligent planning force behind it, and what he said could not be construed as meaning natural selection operating by blind chance. It appears that there is as much blind faith in accepting the evolutionary theory as there exists in some religious dogma.

It would seem then, that the great stumbling blocks of the theory that we evolved from apes are insurmountable, particularly the awesome brain, human speech, creativity and reason, the upright posture and, of course, the total lack of convincing fossil clues to clinch the argument. When we speak of the obvious over-endowment of the brain of mankind seriously challenging the laws of human evolution and natural selection, which does not over-endow a species, we are talking about the average man in the street, but what of the human geniuses that appear from time to time? The child prodigies, the Einsteins, the 200 IQ achievers. They are not just over-endowed but *massively* over-endowed, and highlight even more the tremendous speed of the development of intelligence in mankind in such a comparatively short space of time, and the most unlikely assumption that the human brain is a bequest from any simian type of being.

It seems inconceivable that the slow, plodding evolutionary processes could have handled the massive increase of practically doubling man's brain material in cubic capacity from the earliest hominids, who reputedly had some 500–600cc to the 1,000cc of the *Homo erectus* types of some 600,000 years ago, supposedly the first 'true' human types of the *Homo* genus.

Now, of course, modern man's brain capacity is some 1,300cc, and this highlights an 'oddity' of an actual *decrease* in brain capacity within the general increase. When we go back to Neanderthal man, we find a large 1,400–1,600cc, then settling back somewhat to around 1,350cc for Cro-Magnon man. How can evolutionary processes jump about like this? It is all grist for the mill of the supporters of the 'hybrid man' theory created by an extraterrestrial experimentation programme.

Furthermore, there was a noticeable difference in brain capacity between the hominid types that lived alongside each other, A. Africanus and A. Robustus, for example. The oft-quoted reason for the rapid development of intelligence in man by coming down from the trees, standing upright and freeing his hands for tool making does not seem to be the full answer when we consider dinosaur types existing for millions of years that were fully bipedal, such as for example, Tyrannosaurus Rex, who certainly showed no signs in his upper limbs of manual dexterity capability developing, and Tyrannosaurus Rex lasted for 20 million years. The better equipped kangaroo also showed no signs of utilising hand-like appendages, and, as for the many different types among the ape variety, *their* limit seems to be the picking up of sticks to poke about in ant nests. Man's ability to throw is also quoted as a factor of human development. Chimps mischievously throw mud at visitors in the zoo, but after 40 million years they have not yet started to manufacture any spears to throw at us. They are not even possessed with enough intellect to peel a banana – they split the pod open and scoop the fruit out.

Yet for all the aforementioned and notable increase in the brain capacity, archaeological evidence does not prove that it was utilised much in those early times. It could be likened to a fantastic computer slowly being constructed, modified and improved, but awaiting the final 'switching on', as Cro-Magnon man who certainly had high intellect which was particularly evident in his cave art, *did* exist rather a long time before the obvious intellect he possessed became evident. However, if he was a 'genetically produced' being, this may have been precisely the plan, as he had to get the world in shape first by all that 'tilling and cultivating' before he moved onto higher pursuits. The biblical account in Genesis for human creation has the Creator(s) giving out explicit instructions to mankind to '*subdue and replenish the earth*'.

Cro-Magnon man existed some 30–50,000 years ago, yet he seems to have waited till around 10,000 BC to make his intellectual 'debut'. However, significant archaeological discoveries have been made of Cro-Magnon achievements before 10,000 BC. As he advanced through the millennia, we learn of 'initiators' and 'teachers' in all those legends. Was this the time when the extraterrestrials decided it was time for their first lesson?

If the ethical restraints on genetic experimentation are relaxed in some future time, then just such actions as form the cornerstone of the 'extraterrestrial advanced experiments' theory, will undoubtedly begin. If that becomes the case, it would appear that we will be simply following some racial memory of doing that same process that was done to us, or just following the vast ancient long-term plan for the furtherance of intelligence throughout the universe. To be sure, early man did not need this aforementioned massive over-endowment just to survive. Apes existed, as said, 40–50 million years ago, and have got along and survived to the present day with their abilities to grub about, climb trees, pick fruit, and scratch about in insect mounds. It seems strange that out of all the thousands, perhaps millions of species that inhabit the Earth, man alone qualifies for this excessively high intellect, almost as though we were picked out and groomed for it. Either that or we are just an evolutionary 'mistake', which no one likes to contemplate, or alternatively we are *divinely created*. Fortunately, due to earlier surgeons who threw the ethical rule books out of the window and worked on cadavers brought to them by very shady people, and present day legal brain-donations, we know more about the brain that we would otherwise. But there is still so much to learn.

We talk of being so over-endowed, yet it would appear we only use a certain portion of this fantastic organ. It is apparent in its physical appearance that the important part or forebrain, the part of the cortex that governs our higher functions, *is* highly developed in comparison to other creatures and makes up most of the mass of the brain. This is the part that raises so many awkward questions with regard to its lightning development in comparison to the normal pace of evolutionary processes. If any genetic tampering or enhancement *has* taken place by hypothesised 'ancient astronauts', this is the area in which it would seem that they concentrated their attention upon.

It must be said that evolution and natural selection have produced dramatic changes with regard to other creatures over a relatively short time scale. One example is the horse. Once the size of a large dog, it has changed to its present size, although this is change *within* kind.

Given the ever-increasing amount of molecular groups that are commonly known to us as comprising life's building blocks which are floating in space, and the massive number of host planets we assume must exist containing the right liquid mediums and weather systems to produce lightning and eventually the basic amino acids, then there is no good reason to suppose the process that we think happened here after planetary formation, did not happen on these other worlds.

Of course, exobiologists can offer alternative assemblies of life forms, perhaps for example based on silicone instead of carbon, and referring back again to our obvious bodily vulnerability, there may be far better bodily arrangements abounding on those hypothetical worlds.

In the final section that I call '*Are* Humans 'Children of the Stars'?', and where it is theorised that our world could have been seeded with the so-called blue algae, it is difficult to deride the suggestion that extraterrestrials could have done this when *we* envisage futuristic activity in this region by humans in some, perhaps not too distant, time, on favourable worlds within our system.

The remains of the bacteriological material, i.e. blue-green algae, which our geologists find in ancient rocks, dates back some 3,100 billion years and, although we consider this material basic and primitive, it really is a quite complicated micro-organism. It must have been subject to a long evolutionary process of millions of years. One must ask if that process occurred here on Earth, given the extreme conditions that must have been in existence here on our fairly recently formed world. Furthermore, the sheer volume of the material makes it doubtful that it all came to Earth by meteoric bombardment.

Meteoric analysis is a comparatively new science and is bedevilled by a lack of good fresh samples to be analysed before contamination by Earthly organisms, but it is strongly suggested, if not proven, that meteorites and possibly comets *do* carry amino acids that make up the protein in living cells.

Certain bacterial living organisms may find the environment of deep space positively comfortable compared to some of the conditions in which it has been proved that they can survive here on Earth. There is a species of algae that can survive in hot sulphuric acid and others that can actually reproduce and live normally in boiling water. Of course, bacteria can also survive at the other extreme of the temperature scale.

Micro-organisms have been proved to survive in space as man has actually sent them there and, after they had been subjected to a temperature just a fraction above absolute zero (-272°C), they were found to return to normal when thawed out.

Incredibly, there are bacteria that can survive inside nuclear reactors, but oddly enough can be killed by much less intense radiation, such as the ultraviolet light from the sunlight that falls on all of us. Marine creatures can live in the deepest recesses of our oceans and, because of their adaptation to the enormous pressure there, would blow apart if brought to the surface, much like an astronaut fired into space without his space suit in an unpressurised capsule.

With regard to the whole chain of events from these ancient low life forms to eventually the appearance of man, it seems amazing that we can study life forms that existed more than 3 billion years ago and imagine ancient lightning bolts stirring the organic soup to get things going along the road. We can even reproduce the process in laboratory experiments, but we cannot find the vital links that should be in the ground, showing our evolution in comparatively recent geological time periods.

Of course, the fossil records do show the sudden disappearance of many plant and animal species after just a few thousand years being taken to complete them. There are many such disappearances indicated in the record, the most well-known of them being the dinosaurs' demise, but this is dwarfed by the massive extinction of 90 per cent of all species that seems to have occurred some 245 million years ago.

One rather dramatic explanation suggested for these events is that the sun may have a possible companion star. In other words, we are part of a binary pair of suns with a huge elliptical orbiting distance between them, but every umpteen million years or so the companion shows up to disturb the cometary swarm moving around beyond Pluto and sends a number of them hurtling towards us.

This hypothetical death star has even been named as Nemesis, which is a Greek word describing the goddess of retribution. However, during the sixty or more revolutions of the galaxy from 15 billion years ago, some stars must have come nearer to us than our current stellar neighbour of just over four light years away.

Many would reject out of hand the life-by-accident theory simply by reflecting on the sheer determination of the 'life force' to survive, and the order and intelligence all around us. One example would be to quote some of the many and varied ways that plant life ingeniously disperses its seeds and pollen from the well-known dandelion clock to the bucket orchid, which lures and entraps bees by its nectar pool into which the bees fall, and their only escape is through a pollen-lined tunnel. The flower of the bee orchid resembles a female bee, and when the male bee settles on this to mate, the pollen is assured of being transported away from one flower to the next.

There are many such 'intelligent' ploys adopted by plants in their determination to survive and reproduce themselves. But as insistent a force that life is, the cosmos houses greater forces and dangers evident in the aforementioned wipe-outs where (in the case of the dinosaur extinction) large amounts of iridium are found in the rocks of that geological time.

This iridium, mostly found in meteors, would certainly seem to point to some cosmic visitor, meteoric or cometary, as being responsible for their demise.

To return to the hypothesis of whether life is an accident and serves no purpose, one would have to further define it and decide whether one means life 'per se', or life as we know it on Earth. Of course, we do not know if life exists anywhere other than on Earth and by that I mean the complete jigsaw and not just the pieces that we know permeate space. If we decide we mean life on Earth, then we are saying that the life forms that exist on Earth are the only ones that exist in the entire universe, or we would have to concede that there have been many such accidents on the possible billions of planets (we presume) exist, some of which, surely, are suitable for the process to have commenced given that our aforementioned prerequisites exist there, i.e. liquid, molecular groups and turbulent weather systems.

One would expect that if life was merely an accident, then there would be more chaos and less order, more genetic freaks and more unclassifiable life forms. Instead of clearly defined and formed sea creatures, one might just encounter a heaving, slithering mass, devouring all protein sources in its path, rather like the beings depicted in the early science fiction 'B' movies.

Of course, the opposite is true. There is order, logic, planning and strategy everywhere, from the cosmic celestial mechanics to the detail and planning of everyday mundane chores or those of our profession.

The human brain craves order – in fact it insists on it, which is why we sometimes perceive and construct symmetry from a chaotic jumble where none may exist, such as 'structures' on the moon. To be sure, it is perfectly possible that life, which ultimately forged ahead in great evolutionary leaps to produce an intellect and a great creative brain, *was* a type of genetic accident somewhere out in space, perhaps billions of years ago and then used its evolved intelligence to create order out of chaos. Then, ultimately, it developed space-travel capability and soon began to traverse space, spurring on or encouraging intelligence wherever they encountered suitable life forms to experiment with.

We might consider a life form that began and developed through the enormous time period of the dinosaur epoch here on Earth. Those beings would have had over 200 million years of evolutionary time behind them, and if their development had been as rapid as humanity has been they would be *very* advanced beings by now.

With regard to human development, the theory that we evolved from apes is, at its best, tenuous. There just doesn't seem to be sufficient evidence. Every time one 'expert' in the field claims to have found an obvious link, another will refute it. The theory that we were once apes is quite naturally seen as a little insulting to a lot of people, but one day, perhaps, they will have to grin and bear it if the positive links *are* found; but that time has not yet come. The bone fossils so far found and reputed to be representative of the links between man and the apes do not by any means produce the complete picture of human development from the primates.

One supposes that early man lived alongside the apes in distant times. There were not a threat to each other. The ape would have had quite a few advantages over man, its agile capabilities to reach the highest tree branches for food for example. Bones of humans and apes *would* disappear into the ground together and were later possibly scratched out and pulled about by predators, eventually being found in modern times jumbled together, scientifically dated together, and perhaps winding up on an assembly table together. When the 'Taung Baby' was found in 1924 (a hominid child) fossils of fifteen other different creatures were also found. I certainly would not wish to detract from the experts' knowledge in being able to distinguish between human and simian bone fossils, but there is an inborn desperation to prove the theory with successive qualified people hoping they will be the one that finds the irrefutable proof or 'missing links'. We could again mention the fraud perpetrated in the early part of the century, the so-called 'Piltdown man' named after an area in Sussex where artificially aged bone fossils were discovered as allegedly the much sought-after 'missing link'. We may ask, where will those people who rejected religious doctrines in favour Darwinian theories turn if the theory *does* fall down? They simply *must* find the evidence (or be tempted to create it).

Perhaps the evidence lies buried forever. There is much geological evidence of mayhem and upheaval in Earth's turbulent past. The evidence may lie in high mountains or be sunk beneath a sea bed, but it seems a little too pat, a little too convenient, to assume this reason for the lack of fossil links for *all* creatures, not just mankind. As said, we *did* easily find the dinosaur fossils, in spite of past earthly geological upheaval and it would be stretching coincidence enormously for links showing creatures changing into a different species to also be *all* lost.

Perhaps the creatures of the world did evolve the way the theory is written. Whether human beings fit into this category is somewhat debatable. One thing evolutionary theory needs is time, and lots of it, and it would seem that there hasn't been anywhere like enough time to produce a cellular construction like the human brain. The Earth simply isn't old enough, particularly when we can find a fish such as the coelacanth that looked *just the same* nearly *half a billion* years ago. Clearly, genetic change *is* excruciatingly slow.

As I pointed out in the introductory pages, the human brain was the reason for the aforementioned Mr Wallace's diversion from, and eventual opposition to, the theory. Therefore, it is clear that right from the onset there have been serious misgivings in regard to certain anomalies, such as the human mental evolution, and the lightning development of the human intellectual capacity with still more brain cellular material *in reserve*.

It is mainly the human brain and its capabilities that gives this feeling of insult, not only to people with religious background, but to many others, particularly when they perceive the blank, wooden, unintelligent stare of a chimpanzee from its cage in our local zoos. We are repeatedly bombarded with statements of anthropological 'propaganda', such *as* 'our cousins the chimps' and, 'our ancestors the primates', when the likelihood of the human brain being a simian bequest seems to be non-existent.

So how could this organised mass of reasoning, creative, intelligent, cellular structure, with still more of it presumably yet to develop, come about in such comparatively short time, one could say a blink of an eye in cosmic terms? It seems to clearly signal 'evolution elsewhere' before coming to Earth, giving rise to all the ancient astronaut theories, with extraterrestrial help or enhancement explaining the sudden emergence of intellect, and propounding the theory that our brain is a bequest from beings of another world.

It is clear that changes *have* occurred within species in the past, usually for adaptation or improving survival chances, and this fact probably gave rise to the whole evolutionary theory, coupled with natural selection and survival of the fittest. But change within species must not be confused with one species changing into another. As previously said, the horse was at one time much smaller than it is today, but the fact remains it is still a horse. As the horse has assisted humans so greatly in farming and for transport, perhaps it *was* genetically enlarged for human use by the same hypothetical 'creators'.

Then, of course, there is the chromosome count which determines a species and prevents cross-mating which, as we know, is why man-apes could not be produced by a mere physical insemination with human genetic material. But evolutionary writings will say this *must* have happened and that *must* have happened, the chromosome count must have changed, and so forth. It must be extremely frustrating for those committed to the theory, for there is a lot of musts, maybes and conjecture. A British scientist highlighted around *800* phrases in Darwin's 'Origins' in the subjunctive, such as 'let us assume', or 'we may well suppose' etc.

Perhaps one day all the hard work and diligent dedication to the digging and sifting will pay off, at least with regard to human evolution. One cannot dig everywhere and most efforts seem to centre on Africa as being the birthplace of the dawn of man, but perhaps it will be elsewhere, an area least expected. However, the fact remains that mankind does not fit easily into the theory of evolution and natural selection, but could be seen as a special being, or even a 'misfit' in the concept. True man seems to have appeared just as quickly and mysteriously as the earliest host of creatures did in the so called 'Cambrian explosion' of life forms.

This short span of human history is, I feel quite understated and puzzling but then, as said, perhaps the bones will one day be found. But as it stands today, the true human body shape seems only to have a maximum of 50,000 years of development since Cro-Magnon man's arrival.

It could be said that the only resemblance we have to an ape is if one was seen from a distance temporarily standing upright, which it cannot do for long of course because its hip joints are more suitable for quadruped motion, and not an upright stance.

Of course, there are many other major differences between humans and apes as well as the chromosome count and these will be mentioned in this work. It does indeed seem strange that the beings that supposedly superseded the ape, allegedly on the way to becoming human, *all died* out while apes still exist.

Why is there this seemingly urgent need to prove we came from them? Is it because we and apes are the only tenuous link in the whole process where, it is alleged, creatures turned into something else?

I asked the question, 'Who told evolution to stop?' All the creatures on Earth should theoretically be showing signs of changing into something else, something better, perhaps a more advanced version of their former selves, and if it hasn't stopped one would expect to see half-formed stages where one could compare a creature to its former, perhaps less endowed, version. Where are the half evolved ape men? When were all the creatures issued with their chromosomes and informed by nature to remain as they were? The theory of lightning bolts stoking the primitive soup of the Earth's early oceans is well known, taught and discussed, and it seem to indicate that life, given certain prerequisites, is inevitable throughout the universe.

This oceanic primitive soup, being continually bombarded by high-powered discharges, has of course been reproduced in the laboratory, initially by a well-known experiment in the fifties by a certain Dr Stanley Millar. Although a few amino acids were produced, it is unlikely that anything would have ultimately crawled out of his retort. But it was a start and it seems fairly convincing that life got going that way. But it still had to bridge an enormous gulf from this very basic form to cellular reproduction (not to the mention the right type and number of amino acids required).

Perhaps when we begin to seriously farm the seabeds in some response to a future food crisis, then the oceans responsible in our theories for emergence of life may give up some of their secrets with regard to our mysterious planetary history, producing bone fossils and links of all kinds, and perhaps some 'between species' skeletal forms.

It cannot be disputed that the Earth has had a turbulent past just as surely as any other world has, and allegedly evidence of marine life exists even in the Himalayas. Of course, the process still continues today, and occasionally islands appear and disappear. Coastlines are rising and others are dipping; land masses are still 'breathing out' after the weight of melted ice from the last age retreated.

On these virgin islands a few birds alight for rest and the seeds from their beaks still there from their last mainland pickings drop into the newly appeared mass as they pick at the marine deposits; and the process gets going again.

When one looks at the great divergence, profundity and almost inevitability of life on Earth, in comparison to dead, barren, cold or gaseous worlds that make up our, and probably other systems, one realises the importance of the role of our star and its heat together with our position in the habitability zone or exosphere, yet that same heat or radiation would also terminate our and other life forms very quickly were it not for the protective qualities of our atmosphere which we seem lately to be busily depleting. Although Africa seems to be the favoured venue for human origins, what of the far less obvious areas?

I'm not aware that one has ever been carried out or even envisaged, or could be funded, but perhaps a dig in the Himalayas may equally give up some interesting items to fill in the gaps of our mysterious beginnings. No doubt during this vast period of 4.5 billion years, the mountains and ocean beds *have* shifted quite a few times. We may find the life forms we never knew existed and which may be difficult to classify.

However, trying to establish that we evolved in a natural procession from apes seems to be a daunting task. Perhaps one could more readily believe it if one occasionally saw an ape trying to pick the lock on his cage or trying to imitate speech. How strange that this is left to certain birds, when our alleged closest relative neither attempts the task or has been endowed with the necessary equipment to do so.

Yet when we consider various hominid types that have come and gone in this jerky procession through time, why *have* they disappeared while the apes still remain with us? Also, Neanderthal man's lack of intellect should not (unless purposefully intended) have been a prerequisite for his demise. The apes are not exactly over-endowed in that field themselves, yet here they all are still jumping about in our zoos.

The fossil evidence of human development could indeed begin to look as though it was a sort of 'controlled process' with regard to humans. However, in spite of the obvious over-endowment of intellect in humans, there is a marked contrast in the *rest* of our anatomy, and the human body is a remarkably vulnerable carriage for such a seemingly advanced intelligence.

In evolutionary parlance, we hear of 'survival of the fittest'. How on earth did mankind survive to proliferate as it has? It seems nothing short of a miracle that we, the dominant species who could, or can, or even *do* control the destinies of all other creatures on the planet, are ourselves so utterly helpless and dependent at birth and totally unable to fend for ourselves.

Pigs, cows, sheep and many creatures devoid of creative intellect need no help at all in finding the life-sustaining source of liquid after their birth. Such creatures of the field are on their feet in no time and feeding very shortly afterwards. We, on the other hand, other than perhaps vocally voicing our needs with an indistinguishable noise, seem completely disinterested in finding the life-sustaining nipple and totally lack the muscle power, concentration and co-ordination to get ourselves to it. We have to be lifted to the correct position and have it put into our mouths, or we would starve and expire.

Surely this in itself, if we think about it, suggests an elsewhere evolution aeons of time ahead of Earthly animals. A Utopian existence perhaps, where droids and robotic constructions administer to every whim while the parents continue their intellectual pursuits, unimpeded by the demands of motherhood (or fatherhood), whilst computerised instructional and educational machines commence the process of programming, feeding and caring for the infant. If the *human race* suddenly adopted the practices of the animals and just got on with life after producing their 'mancubs' it would *cease to exist*.

Perhaps a few billion years ago, the early human infant had to hang on to its mother's back while she foraged for food, but many millions of years of evolution rendered it unnecessary. The analogy could be used where a child of immensely wealthy parents with nannies and tutors, bathed, clothed and administered to, suddenly becoming lost in an area of abject poverty with a tough, streetwise young urchin, who has grubbed around begging for food and pennies. Who would be likely to fare better in such a situation? Survival chances would be severely prejudiced by a previous reliance on other sources of support where one had never learned to fend for oneself in precarious situations.

This serves to prove the point that the more advanced and reliant on technology we become, the more helpless we would be if it is suddenly taken away from us, perhaps in the aforementioned global nuclear war or even the departure from a threatened world and the arrival on a new one. Massive degeneration and reversion to savagery would occur, *but* the developed intellect and racial memories would be retained and eventually our past history would no doubt repeat itself, and this is the situation we may be in today, in having this sense of destiny with the stars or the unconscious desire to return to where we may have originated. Our preoccupation or desire to enter deep space seems to suggest either that these memories are 'implanted' as a bequest of alien beings resulting in our unexplainable advanced brain, or that *we* ourselves are the descendants of cosmic ancestors who came to Earth as refugees long in the past. The 'spoilt' nature of humans with this apparent taken-for-granted reliance that something or someone will administer to our whims does not seem to manifest itself in feeding traits only: we lie back and enjoy the privileged period of being spoon-fed for up to between one and one and a half years; we take a further couple of years of leisure before we decide to take on the task of dressing ourselves. And so this weak, strangely vulnerable and flimsy body, which could be seen as inefficiency in a manufactured or designed and created product could be equally seen as another example of an evolution in an environment so advanced that all threats and ills, manual labouring tasks and general wear and tear have been removed, and this is a situation we are working towards on Earth today with our advancements in robotics.

Creatures existing in a 'dog eat dog' environment, or a 'fish eat fish' environment in the case of certain crabs, lobsters and other armed sea creatures, seem well endowed with protective devices, such as the skeleton on the outside. One would have thought that, if we *had* had our total evolution on Earth, we would have over-endowments in our body form to equal those that manifest themselves in our brains.

If we (with our obvious mental advantages) were equipped like the aforementioned creatures, yet still retained manual dexterity and bipedal motion, we could have evolved into something like the creature 'Alien' in the science fiction-cum-horror movie. Since we are quite capable of inflicting enough damage on each other and the planet as we *now* are, perhaps its just as well that we haven't, as we do seem to have retained a dark side to our nature.

These pointers could suggest to supporters of the extraterrestrial 'tampering' theory, that we may be a kind of hybrid, and while retaining the genes of an early simian donor in our savage portion, we also seem endowed with the obvious attributes of a *very advanced planetary race*.

However, the negative forces in the brain do seem ever-present. Why didn't human knowledge and advancement simply continue in an upward mode from, for example, the ancient Greeks? Just what is it that is holding us back? As said, in the great span of time since the dinosaurs' extinction, human development from the earliest emergence we theorise could have occurred twenty times, one after the other. Although we state 2–3 million years, with regard to human development this is being somewhat generous as the first alleged human type, that is *Homo erectus*, did not appear until a time period of up to a million years ago.

If we go back over 100 million years, before the earliest alleged (but now disputed) hominids of 14 million years ago, we find species existing in the same form as they are in today. How could humans come on the scene so quickly and complete their evolution, including upright stature, the, correct skull shape and this enormously over-endowed brain, in such a seemingly ridiculously short space of time?

The current evidence of human development, if anything, encourages the beliefs of those who support the 'genetic experimentation' thesis by unearthly beings. It is easy to see how they can proclaim man is not of the Earth when he has left so little evidence of it. There are many attributes that mankind possesses that could not have been inherited from simian types, such as the ability to design and create and our powers of abstract thought, highly advanced communication, spiritual awareness and religious needs, the sense of ultimate destiny and so forth. There is no chance that the more refined and positive attributes of the human being could have been bequeathed to us by any ape-like creature, a species which has never possessed anything like these qualities *themselves*.

As said, it has been calculated that for the brain of mankind to have developed its 100 billion neurons or brain cells, the length of time required should have been in the region of billions of years of evolution, rather than millions. Not surprisingly, one is tempted to ask if our major brain evolution took place elsewhere in space before the four and half billion years old Earth was even formed. If so, we would have to concede the possibility that our cosmic ancestors arrived here from elsewhere in space, or that our brain is a genetic bequest by unearthly beings using their own material in some form of experimentation with the primates.

When the first fish struggled out of the sea on its way to becoming an amphibian away from the many predators that would devour it, and went through all the processes our drawings and cartoons tell us of the way it happened, finally becoming bipedal and able to use its 'hands' to manipulate things, it did not signify a great advancement in the other creatures already mentioned. This includes all the large ape and primate varieties who have not made any great use of their gift of manual dexterity other than simple actions relative to their environment, and are most uncreative beings in comparison to other animal and insect varieties.

Clearly it *did* with regard to humans, *because of the excess of intelligence*. But bipedal stance with arms and hands, as the evidence shows, does not equal intelligent advancement. At the moment there is no evidence to show that man has ever been anything else than bipedal but, as said, apes have not taken advantage or made any great use of *their* hands in their huge evolutionary time period. However, the much sought-after fossil links *may* one day be found to complete the picture. There never seems to be a shortage of volunteers for the task, and we seem to see a continual and never-ending procession of archaeologists who are quite willing to go out into environmentally hostile areas and happily dig around. If they do not find any old cities or civilisations, they may just find those missing links that have long been searched for.

As well as the time specified for human development being pushed back, the computed age of the universe itself is now seemingly in doubt, and coming down in some estimations towards a single figure, instead of the 20 billion years originally estimated. When we start to view the universe itself as being younger than some of the ages of the stars we have looked at, clearly the sums are going wrong somewhere. One wonders if *any* of our dating processes are reliable and whether they may be flawed in some way. However, it is clear that the bones we find to construct all these impressive dinosaur skeletons are considerably older than the human fossil evidence currently available.

We should today be able to go into any natural history museum and, as well as viewing the dinosaur specimens with every bone in place, also be able to see the complete picture in constructed skeletons of apes gradually becoming humans, with hip joints gradually changing to give the apes upright stance and skulls showing clear and gradual change without all the plaster and cement added to fragments to produce theoretical 'in-between stages'. Human fossil finds are disappointingly few in number.

It certainly is mysterious that we cannot find the bones, and we seem to be heading towards the inescapable conclusion that they are not *there* to *be* found. The meagre collection of fossil finds currently in our possession does not, as said, seem to represent over a century of searching.

People, for the most part generally trust science to eventually come up with the answers to the problems that puzzle us and to answer most of the queries about life's mysteries, but it has to be said that the science of anthropology is knowingly, or unknowingly, misleading a large number of the population by allowing them to assume that, with regard to the 'theory' of evolution and natural selection, it is all proven and all the facts are in. Eminent anthropologists will readily admit that all the facts are most certainly not 'all in'; nevertheless, there are certain statements made that are phrased in such a way as to be misleading.

The misleading (conscious or otherwise) is done in a number of subtle ways. Firstly there is a frequent reference, wherever possible, to the phrase 'our cousins the chimps', or 'our ancestors the primates'. Then there are all those diagrams in zoos, theme parks and 'ancient caverns', showing the early sea creature emerging from the ocean to go through all that shape-shifting of species, changing into something else and gradually to apes, attaining an upright stance, to humans, without a particle of evidence in the form of a skeleton of a 'halfway' stage between one species and something else, and so the long journey along the evolutionary road from lungfish to humans will have to be depicted by drawings rather than skeletons.

It is fairly easy to achieve this propaganda of the masses as those people who *are* religiously orientated ignore the seemingly atheistic and scientifically cold theories of 'evolution' for human appearance, and others with more of a tendency for evolution and natural selection, for the most part do not look at the facts available to support it.

Another method used to attempt to reinforce the theory as fact, is to continually omit the words 'the theory of' and speak only of 'evolution from the primates'. Then there is the aforementioned 'invention' of a species named 'Pro-Consul', a creature resembling the type of being that they would like to construct in reality if the necessary bone fossil links could be found to enable them so to do.

Now it would appear that a much more subtle approach is being attempted through the medium of television. Recently, a TV programme depicted a scenario formulated in the manner of a 'Crown Court trial' to decide on the evidence available and the pros and cons of the need to confer 'human rights' on chimpanzees. Many people would have seen the implications in the programme of once more trying 'through the back door' as it were, to reinforce the 'apes to men' theory as the emphasis was on how chimpanzees are so like us (once again were referred to as 'our cousins') and that they should have human rights.

Great emphasis was made on the percentage of DNA and the blood similarity. Many people who may have had some doubt must have thought, 'If they are seeking human rights for chimps, then perhaps they really *are* our cousins.' It seems abundantly clear that the sheer frustration of the inability to find the convincing proof and clear evidence of the theory and place it on the fossil table produces a kind of 'the end justifies the means' persuasive brainwashing of the masses. One may well ask, in relation to the aforementioned television programme, are we also to have human rights for pigs since *their* bodily tissue contributes to saving human lives in heart operations? What about human rights for parrots and mynah birds, since they can imitate human speech successfully which is something apes or chimps could never do, as they have never had and never will have, the necessary equipment in their throats?

In India they already have a situation where *cows* have a form of human rights, and cars either go around them or respectfully wait for them to move on. Perhaps we should confer human rights on them here and take down all the fences. The whole concept of human rights for animals of any kind is silly in the extreme. It seems strange that human rights for animals can even be considered when a whole industry exists to get some of them to our dinner tables to be eaten.

We have a *responsibility* to ensure that animals are adequately cared for and, although the law requires amending with regard to cruelty and blood sports such as fox hunting and so forth, 'animal rights' should not read 'human rights'.

All creatures have DNA and blood and communication skills of one form or another, and some creatures *are* quite creative. One never sees a chimpanzee build *anything* yet many other creatures build quite complicated environmental habitats. The ants have amazing co-ordination and organisation, even herding aphids like cattle, and could not have such organisation without good communication. Clearly, there are many creatures with high intelligence relative to their functions within their specific environment who seem much more intelligent than our 'cousins' the chimps, and their long evolution has served them well. It has provided them, in some cases, with creative abilities sufficient to survive: high sensory equipment such as in the shark, or extremely good eyesight such as in predatory birds like the hawk, eagle and the owl – indeed in birds in general.

However, when it comes to the chimps and other simians, where one would expect to find such advanced qualities, they are extremely poorly endowed and uncreative. Yet people working with them seem amazed by the performance of a few simple tasks after they have had enough evolutionary existence to be masters of the universe.

After the period of their evolution, humans will have no excuse for not only being masters of the universe, but probably rearranging it. In Section III we will look at the possibility of such beings existing with a far longer evolutionary time than even the age of the Earth itself – the creators not only controlling their own destiny, but possibly that of their creations. We humans may be among that number. Such beings may have seen everything, learned everything, achieved everything. There would be nothing else for them to do but create. They would be likened to 'gods'. They may even have created pulsars as celestial navigation beacons whose intermittent signals are as accurate as atomic clocks.

Nothing would be too difficult for such beings. Did such beings come by the Earth over 3 billion years ago and 'seed' it with the blue-green algae found today in Earth's most ancient rocks?

Section II
'Divine' Human Creation! – Questions

1. In the Beginning

When dealing with creation, it is necessary to be specific about the difference between 'creation' of the Genesis kind and 'scientific creation'. In point of fact there is a very thin line between the two. One would only have to accept that our well-known 'big bang' occurred by divine command. 'Let there be light' would do very nicely. It is certainly hard to envisage matter coming into being from nothing, so where did that huge primordial mass of hydrogen come from? If this hypothesis is accepted as a divine command, one could blend one's view on evolution and natural selection with a religious view, but one would have to view the events in the Bible in the Garden of Eden (which are quite specific) as a myth or a parable about right and wrong – our choice between good and evil. If we reflect on the concept of Adam and Eve, here we have two people of undetermined racial type, alleged to be the progenitors of the entire human race, which would require the production of the three distinctive types that make up the population of the world, i.e. Negroid, Oriental and Caucasian.

Everyone is familiar with the story of the happenings in the Garden of Eden. Some races have their own version of it but the serpent comes over clearly as the source of evil. Strangely, serpent gods and deities abound throughout the historic writings and legends of many lands, and are not in many cases seen as evil but as knowledgeable and helpful to mankind. In a speech to his Apostles, Christ himself said, 'Be ye therefore wise as serpents' (Matthew 10:16).

Legends from San Cristobal Island say that the serpent beings are the 'creators'. The sages of ancient Egypt and Babylon regarded themselves as 'sons of the Serpent God'. Most certainly the symbol of the snake or serpent can be seen all over the world. A legend from the Solomon Islands refers to inter-breeding between serpent deities and Earth people. Old Sumerian, Babylonian, Egyptian and Greek legends refer to the serpent deities who were believed to have once resided in the underworld. Legends from Aboriginal and South Pacific sources speak of these serpent beings again being associated with the creation legends.

The whole concept of the *downfall* of man from the events that allegedly took place in the Garden of Eden by a 'serpent-like' being, the driving quest for knowledge in Adam bringing about his downfall, stems from the thirst for knowledge given to mankind in the first place – by the Creator according to religious belief and by evolutionary processes according to other beliefs. This vastly superior brain ultimately became responsible for the *ascent* of man. Is it possible to believe in the events that allegedly took place in this 'Garden of Eden'. Was the 'serpent' a special being that carried out certain strength of character checks and tests for signs of enquiring intelligence in the human creations and later denigrated through time to a snake?

Certainly, there are many people today who happily accept the biblical Old Testament version of the creation in the Genesis account without question, and proclaim that it occurred exactly as it is written. Well they are entitled to their beliefs. If this gives one a certain contentment and peace of mind, then so be it. However, the problem that arises in so doing is that one would have to avoid at all costs any form of debate and evaluation, any form of discussion or questioning and probing made by others less convinced of the biblical writings, that could cause one who believed so rigidly to perhaps have doubts or second thoughts that would most certainly rock the boat and disturb their formerly tranquil peace of mind.

Many theologians themselves now reject the Genesis account of creation as being true in its entirety exactly as it is written. They will tend to put forward or offer more logical and modern day concepts while still attempting to retain the central theme.

Some will defend or explain the six or seven days mentioned to complete the whole creation event as not actually meaning days, but rather 'eras', or periods of time – just as when an older person may say to a younger one, 'Things were never like that in *my* day', clearly not referring to a period of twenty-four hours.

I think we can say with certainty that creation (that is, the actual production of the material making up the universe and all its contents including us) *is* a fact. By that I mean the stupendous, uncanny explosion alleged to have occurred 15–20 billion years ago according to the patient studies and efforts of numerous scientists, astronomers and astrophysicists. (However the latest estimates are 11 billion years or possibly *less*.)

Now this was indeed a fantastic event, creating the basic particles of all things, throwing them out in all directions to form galaxies and suns like our own and not quite like our own: the hot blue stars hurriedly burning up their energy, and the huge orange stars that in their death throes produce a massive annihilation of themselves in supernovae, flinging the material cooked up in their fiery hearts outwards – material that we ourselves are made of. And so even in their demise they help to 'create', as our very bodies are made of this stellar material.

It is hard to deny that this particular event was in itself a form of creation. During the advance of science, which included the advent of spectrographic analysis of the radiations that make up the visible light all around us and the breaking up of these radiations into their various colour bands, the shifts in the spectrum told the astronomical scientists that most of the galaxies appeared to be hurtling away from us at ever-increasing speeds. Because of this it was quite logical to deduce that we seemed to exist in the midst of the massive 'explosion'.

The next logical step was to compute it all back to a starting point which is our well known 'big bang'. We look on this, of course, as a scientific creation, as an astronomical event and do not usually refer to it or relate to it in religious terms. The biblical account sums up the colossal occurrence just referred to in four words: 'Let there be light', spoken by the Almighty when creating everything in six days. The Creator certainly deserved His rest on the seventh day after completing such an achievement in six days. The 'big bang' almost certainly happened as science decrees. What we have to come to terms with is whether it was a spontaneous, perhaps inevitable event, or was there intelligence behind it? It is certainly very hard to imagine matter coming into existence from nothing. We ask, what was there before it? How did this massive volume of material appear from nowhere? How did it manage to explode at all? Such a volume of material should have been an enormous black hole.

If we assume that the dating processes used to determine the age of fossil finds and rock seams are reasonably accurate, based on the known values of radioactive decay, etc., then we have this figure of 4.5 billion years for the age of the Earth. What on earth is the Genesis version of the event talking about when referring to six days? Given that our computed age for the Earth is correct then, allowing for the few million years for man's appearance, each day works out at being something like 750 million years long. After reading such things in the Old Testament, surely it makes it much harder for people struggling to accept it to actually do so.

I am certainly not criticising religious beliefs. Christianity is a wonderful religion and, if everyone abided by its teachings and commandments, we would not have the wars, crime, evil actions, greed and intolerance that exists today. But it must be said that the Old Testament writers, who must have been intelligent people, seemed to just carry on blissfully with the writings without considering that some of the things they were writing might appear 'questionable' in the extreme to lesser mortals. Adam and Eve, for instance. The Bible teaches that they had two sons, yet they were supposed to have been the progenitors of the entire human race. Surely a strange thing, as one slew the other and, most importantly, there was no further mention of wives being created for them. Even if Adam and Eve *had* produced offspring of different gender, it would have been incestuous in order for the process to continue. Not a very good start. Later in the biblical text, reference is made to Cain's wife, but she is not mentioned as being specifically created in the Bible. In fact no other 'creations' other than Adam and Eve are mentioned.

Long ago, reading (like writing) was the prerogative of the privileged few, the educated ones. Surely the Old Testament writers would know their words would not be read by idiots. What of the masses that received the contents through preaching? Did they not question it? Why was it never amended, never explained? Even to this day it remains in the same form. Moses, the traditional writer of Genesis, was raised and educated in Egypt and he would have possessed a questioning intellect. He must have known that these obvious difficulties presented in the process of starting the human race would be noticed or queried.

The idea of some wondrous Being coming along and then carrying out these fantastic acts is, to many people, totally unbelievable, *however long* He took. But to accomplish it all in six days?! In order to write the books of Genesis, Moses must have unquestioningly accepted it.

However, as will be shown, the distinct possibility exists that the biblical account of Genesis attributed to Moses may not be in the *original* form in which it was written.

Surely, if the fantastic Being, the Creator, was capable of such tremendous feats as creating an entire universe and everything in it in such a short period of time, that Being would most certainly not waste time breathing into lumps of clay and removing ribs. He would simply just wave an airy hand and proclaim, 'Let there be people'. The removal of genetic material from Adam is more suggestive of some kind of 'cloning' process, or 'genetic creation'.

The so called 'big bang', we must admit, was indeed creation and where there is creation one must surely assume the existence of a creator, but the way in which

Genesis is written causes many people to struggle to come to terms with it.

Suppose the original version written by Moses did contain some references to a more scientific type of process. It would almost certainly have been altered by religiously zealous popes in the early Middle Ages to make it more divine in nature and rid it of the 'satanic science'. The author Robert Charroux in 'Lost Worlds' (Fontana) states that many alterations of the Gospels from the early popes down to the invention of the printing press (which effectively put a stop to it) *did* occur.

In a recently completed book that I have called *The Angels of Abraham*, I have carried out an in-depth study and analysis of the 'close encounters' in the Old Testament that the patriarchs such as Abraham and Moses had with beings they interpreted as being 'divine angels'. Anyone who has read the Old Testament in detail will know that there are a fantastic amount of them and, if we accept the existence of the patriarchs and that they were truthful men, these interactions with 'angels' were very profound. The angels descended and ascended on pillars of fire, conversed, ate, drank and 'directed' their chosen beings (the patriarchs) to do their bidding. Moses was an intelligent man with a superhuman memory, or an extremely efficient filing index or data source. This is evident in his enormous list of characters in the Old Testament running through many generations, of who 'begat' who, their locations, offspring and how long they lived and so forth. Moses must have made intelligent enquiry regarding things that may have puzzled him during his frequent encounters with the 'angels'. Consider the following logical assumptions and deductions.

Moses was allegedly raised and educated in Egypt on a par with the Pharaoh (Ramses II?) and would have had access to the great libraries, repositories and learning centres that contained amazing scrolls, parchments and documents that clearly enlightened the Greeks when they overran that land only a few centuries after Moses. Unlike subsequent conquerors, such as Julius Caesar and the Caliph Omar who sacked, pillaged and burned the great libraries and their works, the Greeks obviously *learned* from them and made amazing statements that we now know to be accurate. Anaximenes spoke of the remoteness of the stars *and their non-luminous companions.* Anaxagoras stated that 'other worlds' existed. Democritus knew that the Milky Way was an immense multitude of stars and that everything was atoms and space. Pythagoras knew that the Earth was a sphere. Aristarchus knew that the Earth revolved around the sun. Anaximander theorised that all species evolved from a common source *over 2,000 years before Darwin*.

All these men, without exception, would have been burned at the stake many centuries later in the dark unenlightened Middle Ages. Moses, when writing the Exodus story of his own exploits in leading the Hebrews, compels one to accept the extraterrestrial hypothesis regarding the actions of the angels, which in many cases were far from the description 'divine'.

In the cool of his tent, when conversing with the 'angels', let us assume Moses asked of them, 'Where exactly is Heaven?' Suppose their reply was, 'We travelled many light years across space to reach your world over 50,000 of your years ago. We are responsible for your very existence. You originated in our life-creation centre we called 'Eden'. The 'angels', who I conclude in *The Angels of Abraham* were most certainly extraterrestrial beings, would have explained human creation to Moses in terms he would relate to. If these enlightenments had been mentioned in Moses' original account of Genesis they would have shocked and horrified medieval religious zealots, but not the earlier people of Earth who listened and pondered on the teachings of the ancient Greeks whose sources of knowledge may have been quite familiar to Moses.

Just like the urgent motivating force that caused the anthropological fraud of the 'Piltdown Man' in the desperation to have the apes-to-men theories accepted, the religious zealots also resorted to fraud with forgeries of religious icons and artefacts such as saintly relics and the now disputed Turin Shroud. Just as many people today would wish the entire biblical teachings of the Old Testament to be true in every detail, even the fantastic Genesis account of creation, our logical processes make it difficult for us to accept certain accounts in their present form, particularly human creation. With the obvious knowledge and enlightenment possessed by Moses, the possibility exists that the Genesis account of human creation may have been quite different in its *original* form. Our acceptance of the fantastic explosive 'happening' of 15–20 billion years ago seems to be universally agreed.

But was there a purposeful 'intelligence' behind that colossal event? When one looks at the ultimate results, that is all the galaxies, suns and (presumably) planetary systems as well as all the creatures of the Earth, including us, it would certainly convince many that there was.

Many bright youngsters with enquiring minds, when questioning certain alleged events in religious teachings during Religious Education lessons, may be quite unsatisfied with the usual and expected reply that they must 'have faith'. The teachers may be reluctant, or afraid somehow, to just say, 'I don't know', 'I can't explain it', 'I'm concerned about it myself', 'I'd like to find out'. There was always this 'blind' faith. They probably would be told the parable of doubting Thomas, who denied Christ had risen until he could actually see and feel the wounds, and they would be made to feel thoroughly guilty for even questioning things. This situation certainly prevailed in cloistered Catholic schools during my youth.

It is not surprising that, as children grow older, they move away from such dogmatic teachings when learning to think for themselves. Religious teachings pronounce that God, the Creator, *is*, always *was* and always *will be.* Now, compared with the quite unimaginable length of time that is eternity, 15 billion years comes more into perspective and does not seem to be, in relation to eternity, such a tremendously long time after all. So we must now ask: what was the Creator doing before the main event, his major production? Will it ever be revealed to us what wondrous acts were taking place before this major act of creation? Surely, He must have been occupied in some way during the vast span of time leading up to our famed big bang?

We as humans are also, in our way, quite creative, and anything we create we automatically assess during its construction, critically review it afterwards, then modify, alter or improve on it – perhaps even scrap it altogether. Our whole existence revolves around our supreme creative capabilities from H-bombs in the negative to aid packages in the positive.

So, one must ask the question: is it likely that the Creator is satisfied with His handiwork, i.e. us? Is He completely satisfied, relatively satisfied, or not satisfied at all? If He is completely satisfied, then He is obviously very easily pleased. If only relatively satisfied, or not satisfied at all, this implies some corrective treatment to be carried out shortly or some time further in the future.

Is He wringing his hands, wondering where he went wrong? For, clearly, we are far from perfect created beings. The 'Jekyll and Hyde condition' that is part of the human psyche was long ago recognised and gave rise to the story the same name, the Chinese Yin and Yang: the ever-continuing struggle of good and evil, the negative and positive brain patterns seemingly locked in a never-ending struggle of one trying to dominate the other. Will one side eventually prevail? What if it is the evil side? Perhaps only neurological surgery, or some sort of genetic implant, will be needed to get us over the hump so to speak – that necessary little boost for our brains to develop enough to eliminate the wasteful and negative motivations. If we consider the extraterrestrial hypothesis for human brain development, then 'they' may also see this need for genetic 'correction' which may be seen as a very simple process (to them). However getting humanity to 'co-operate' would be an enormous problem for them.

Most religious teachings have woven it into their doctrines that mankind was purposefully given this choice between good and evil. The decision is ours, as logical reasoning beings, to make our *own* choice. It is all down to *us* and nothing to do with the Creator at all.

He will just make sure that if we make the wrong one we will roast in Hell for all eternity. Clearly, if everyone really believed the doctrines there would not be any evil people. They would be far too frightened to be evil.

It does, however, all seem a bit puzzling. If the Creator had the ability to produce only good, right-minded beings, why did He not do so in the first place instead of giving us this built-in fault that for many will result in their paying for all eternity? Surely a strange thing.

If we spin a globe of the Earth and stop it at any point, then if it is a land mass and not a sea area, the chances are that the country under our finger will have either organised crime, or a soaring crime rate, drugs, prostitution, muggings, rape, murders, or perhaps a handful of dictatorial gangsters, or some junta or other calling the shots, probably spending their nation's resources on the hardware for war and happily engaging in it instead of feeding their people. They can head for the nearest border or aid plane – let the rest of the world feed their people.

As we watch the whole dismal affair on our television sets, usually at mealtimes, we see the children with distended stomachs and flies in their mouths. So we guiltily push our dinner to one side and fill the little envelope that comes plopping through the letter box, no longer asking us to give what we can, but: 'Five pounds will do nicely, thank you'. Not once are the so-called leaders of the countries who are causing or allowing such mayhem to flourish mentioned. We never know their names. The emphasis is always on the do-gooders, charity workers, medical staff, nuns and so on.

Never on what course of action is proposed to get rid of the *cause*, or bring them to justice under the UN mandate for crimes against humanity. If the British can send a task force and retake the Falkland Islands, and if the US can barge into Haiti and tell the military junta to pack their bags as happened in September 1994, we may ask, 'Why can't the same action be taken against these unmentioned 'faceless' ones by the UN?'

So, the Creator sees it all happening before Him. Does He ever reflect on where He may have gone wrong, like a parent of a delinquent child? Of course, He also sees there are many good people, those that will go readily to these troubled areas, and, forgetting all thoughts of danger and disease, or risk to themselves, admirably carry out their tasks.

He also sees great intelligence in His creations, the great thinkers, philosophers, mathematicians, scientists, nuclear physicists; the Niels Bohrs, the Oppenheimers, the Einsteins, the best brains in the world who could have lent their great intellects, along with all the others, to some great think tank dedicated primarily to the creation of some master plan that would accentuate the positive tendencies of the human brain, eliminate wars and disease and work towards the elimination of all the negative and destructive forces in the mind. Instead, they lend their intellects towards the construction of a single device, so terribly powerful it can kill hundreds of thousands of people with a single detonation. Surely this is a strange thing? However, in spite of the equal and opposite *negative* forces lurking within our minds, the human brain is in itself a fantastic creation, an enigma, a mystery. It seems incredible that it could have come into being in such a short period of time.

In fact it *has* been stated that the Earth simply isn't old enough to have produced such an organ by natural processes. As said in Section I, it hardly seems likely at all to be a terrestrially evolved organ, and so we seem to be left with two alternatives: the brain is a gift from *God* or from the 'gods'. The fact remains that whichever we choose, there would be concern in the 'donor' over the negative functions of this otherwise fantastic organ.

There are other creatures on the planet with high intelligence levels, such as the dolphin, which strangely seem to have an affinity with man and quite often befriend and entertain him and are easy to train because of this high intelligence. But creatures like the dolphin and the shark are clearly restricted by their environmental surroundings and are probably as advanced as they need to be for that environment.

Mankind, on the other hand, and as the creation story goes, has 'dominion' over all the other creatures of the world and is now equally at home on the planet's surface and (with survival apparatus) also able to function quite well under water and in space. Did an almighty Creator intend mankind to ultimately strive to reach into *His* domain?

We tend to take our intelligence for granted and seldom stop to think just how superbly equipped we are in comparison to other creatures with our creative and reasoning powers, bipedal locomotion and manual dexterity for manipulation. Although our bodily structure is very unprotected for climatic extremes, or for wear and tear, our brain makes up for it by our intelligence in knowing how to protect ourselves with clothing and avoiding danger areas. But what of the darker side of our nature? Was an almighty Creator responsible for that as well?

Clearly, as well as being quite capable of great feats in positive creative action and ability, we can also plumb the depths of depravity. So can we assume, then, that the Creator is entirely satisfied with His creation? He has seen all the evil that we are capable of – surely this tendency in some of us towards our dark side was not part of the plan, no matter how religion explains it away?

We must consider, however, the existence of all the good people along with the bad. Surely we can expect some selection and separation of certain groups. Will they receive a command from on high to build a modern day ark? What about the rest? Perhaps a destruction like that of Sodom and Gomorrah, a pestilence or a plague of boils. But given the extent of worldwide crime and general evil actions, where would He start?

To be sure, He must be quite capable of such remedies and was not slow in coming forward for such actions in Old Testament times. It is doubtful that we are any more evil today than we were a couple of thousand years ago. It's just that we are more efficient at it with the advance of technology and weaponry. It is certain that if we built a robot city and filled it with active units and they suddenly began to malfunction or even terminate each other, we *their* creators would soon step in remove the offending units, tinker with their intelligence packs, try to correct them, modify their motivating systems and, if beyond repair, possibly eliminate them altogether and maybe start all over again, hoping next time we would get it right. (This kind of scenario was suitably depicted in the entertaining movie *West World* starring Yul Brynner, where lifelike robots malfunctioned.)

We may ask, how long will mankind be allowed to malfunction before the Creator steps in to carry out the necessary repairs and modifications? But even without divine intervention or retribution, there are many other threats to our very existence.

Almost any day we could be all but wiped out by the detached and uninterested violence of the cosmos, some celestial event or threat such as, for example, the arrival of a couple of huge asteroids knocked onto a collision course with Earth by some wayward comet passing through their back yard, or something similar to the broken up comet that (fortunately) rained down on Jupiter instead of Earth.

Perhaps a nearby supernova blowing away our atmosphere or a huge comet tearing into our atmosphere, this time not being so small or not picking out a remote unpopulated tundra region? An event of our own making, such as a runaway greenhouse effect? Our own sun suddenly flaring up and devouring the inner planetary bodies, including Earth and possibly Mars? On this last point, I think we can rest fairly easily, for we seem quite fortunately to have a steady old sun about halfway into its main sequence.

But the aforementioned points serve to illustrate our vulnerability to extinction without any majestic retribution from the Creator and, in any case, these possibilities were designed and built into the system by the Creator Himself when producing it all. Or were they all unforeseen and not part of the plan? If so, this implies an imperfect Creator and this would be unacceptable to many. So their choice would be that they were 'built in'. What delicious macabre humour.

On the point of our own star and its apparent stability in comparison to some other suns, one must assume, if accepting the divine creation theory, that some of the stars must have been produced simply to decorate the night sky for *us* and for *our* enjoyment, as otherwise there would be no point in their existence:

The hot blue stars, burning up their hydrogen at a furious rate and probably not existing long enough to allow life to gain a foothold or eventually get started, on worlds that may exist and occupy favourable positions in their habitability zones.

The huge, massive stars that will destroy any life that *had* got started in their system, and probably also planetary systems within a few light years' distance, when they reach the end of their life and explode in what is termed a 'supernova'.

However, the Creator may really be a bit too busy at the moment to be continually watching and worrying about the crazy antics of mankind. He may have much more troublesome things to worry about. There may be millions of civilisations existing that are the results of His handiwork, and some may have far worse behaviour patterns than we have. They may have a much higher priority on the repair list than us.

Furthermore, there are the 'knock-on' effects of His handiwork to attend to, the ever-continuing events we ourselves can now observe with our radio telescopes so far away: the spontaneous creation of stars from great gas clouds and (most likely) planets forming in their turn afterwards. There is much to oversee and supervise. Man and his petty little problems can wait their turn.

Our lives and our problems do seem small beer when reflecting on our great galaxy, along with others, that seem to be swirling like immense Catherine wheels. Yet the dinosaurs evolved and lasted for some 180 million years, and most of the life forms have evolved on Earth, including our own evolution, during the time of a single revolution.

Even during the revolving of the galaxy, threats to our very existence abound. We may come upon a free moving star, or enter a darker, dusty zone, and be plunged into a severe and long-lasting ice age by the blocking of our life-giving sunlight. Temperatures may drop so low that all life is extinguished on Earth and the seas freeze over.

The universe seems cold and indifferent and totally unaware of our petty existence, but of course we must always 'Look on the bright side of life' as the song goes, and reflect that even though we revolve around the galaxy in our outer arm, the sun and its planets bob in and out of the galactic plane during this long revolution, so we may just avoid the nasty old dust after all.

To return to the *main* creative event again, the so-called 'big bang'. The old theories and arguments as to whether the galaxies will continue their frantic rush away from us and eventually all wink out, or whether their flight will gradually be arrested and slowly decelerate and grind to a halt before beginning an inward fall back to the point of departure once again still seem to prevail. It all revolves around the question of the so-called missing matter that would be required for the gravitational mass necessary to affect the departing galaxies and slow them down, and start the return process.

The steady state or expanding universe? Perhaps, with the advances in astronomical science and radio astronomy, with bigger and better orbiting telescopes, we will soon have all the answers we seek with many burnt out suns accounting for the missing mass. If all that material does come crashing back again, only to form another huge, explosive ball, the process could have been going on for all eternity with life arising and becoming extinguished as regularly as clockwork.

We may all just be caught up in something that has been going on for aeons, each time exploding from the 'primordial egg' of hydrogen, which even at the moment accounts for most of the mass of the universe. *That* continual explosive creation would be like existing on an electron, circling an atom that helps to make up the bodily structure of some vast breathing being. We, on the planet, may just be a virus, simply attacking a bodily cell.

It is strange that we can watch all the wonders of creation through radio telescopes actually occurring far away, and know so much about the workings of the universe, yet we do not yet know for sure that planets even exist in orbit around all those suns, apart from some flimsy evidence of perturbation effects by the gravitational attraction of nearby bodies (which may even be small compact stars).

Of course, it seems ludicrous to think that a Creator would produce just one liveable planet among the nine worlds our sun possesses, and that no others exist anywhere else in the entire universe. Ours is a rather mediocre old sun and there are others like it, but 'single' stars seem to be a rarity in comparison to binaries and triple star systems. If planets do exist in those systems they must be very unstable and erratic in their orbits, especially if the solar orbiting bodies are fairly close to each other.

It is strange that there is no mention of other extraterrestrial civilisations existing in all that creation in the biblical teachings. The Bible certainly allows one to assume that our world is the only one in existence.

It seems certain that if there had been teachings and references to other-world populations in the heavens, it would have been accepted in those times without too much trauma and, of course, by now we would all be well used to the idea. If this had been mentioned, even though the Genesis account does read somewhat like an imaginative tale, the main body or gist of it may well have been accepted by many more people, but that it has all been done just for our benefit cannot be accepted by many.

It also seems certain that the discovery of extraterrestrial intelligence today would cause more trauma, cultural shock and social disorientation than it would have done in former times, because after the writings had been set down there were plenty of marauding armies sweeping across continents to occupy, subdue, enslave and to rape and pillage. The people of those times had closer-to-home things to occupy and worry them than any extraterrestrial threat.

If we believe in a Creator, surely we must assume that He produced life on at least some of those other possible worlds, and if He did, then why didn't He mention it? As said, it doesn't seem likely there would have been worldwide panic. It would have seemed quite natural to believers of the old Genesis account. Of course, He did say He was 'not of this Earth'. Perhaps His real home is planet Heaven, somewhere in the galaxy we may one day know of.

If we do accept each 'day' in the creation story as being an era and not twenty four hours, then that era, as aforesaid, amounts to about 750 million years, so quite a lot could happen with regard to the formation of the Earth in one of those 'days'.

When we look back into the far distant past, geologically we find that something interesting was going on around one of these 'days'. Today we hear the word 'terraform' mentioned and astronomical science and astrophysicists talk of bringing otherwise dead worlds to life by terraforming processes, which basically require the introduction of masses of blue-green algae to take hold and start the long process of converting a planet of a suitable type into a liveable world for our life forms. Interestingly, Russian astrophysicists have proposed that the atmosphere of Venus could be conditioned for our life forms within 600 years.

If we go back far enough in our own time, we find just this substance in our own ancient rocks which begs the question: was this process intelligently introduced to *our* world long ago to make it liveable for us, either by divine methods or possibly by our ancient outer space ancestors travelling through space with a duty to create or 'enhance' life forms wherever possible? Perhaps the 'enhancing' on Earth took place with the Neanderthals to produce Cro-Magnon man, our real ancient ancestor as suggested herein.

It seems certain *we* will carry out these types of activities in the future. Perhaps the actions of forming suitable worlds for ourselves are all part of the original instructions. Perhaps in the future, we will be happily multiplying well beyond the confines of our own world. The biblical instruction to 'go forth and multiply' may not have been meant only in regard to our Earthly environment.

With regard to the creation of the universe, whether it is a divine or natural event, can we really be self-centred enough to assume it was all brought about with us in mind, that we are the centre of things here in our probably mundane planetary system, orbiting an average star somewhere on the outskirts of the galactic arm?

If not, then does the creation story apply equally to all the other Earth-like planets that surely must exist by sheer weight of (computed) numbers? Are those hypothetical beings also encouraged to 'go forth and multiply' and to aim to eventually reach Heaven?

Heaven must have a very severe population problem if all these other (though still mythical) beings go there – or are there many different Heavens to cater for all the other deceased beings from all those other worlds, not only in our galaxy, but in the countless others in the universe?

So perhaps we *are* the first, and it *is* all down to us to spread our life forms throughout the galaxy and, maybe, eventually the entire universe. If we do 'seed' other planets, Mars or Venus perhaps, and then much further into the future, populate them, then the process will have started. But as said, there is nothing spectacular about our mundane star and its position in the galaxy. Older systems must exist, so why us?

Furthermore, with our continuing wars and savagery towards our fellow men, we still have not got our own house in order, let alone made plans to condition and populate other worlds.

We have all these grandiose plans of space stations, deep space voyages and bases on other worlds. We send and wait to receive messages from other worlds, contemplate the existence of life on other worlds and envisage trips there, when sometimes it seems our *own* future is in serious doubt, right here on Earth, unless we restrain our ecologically damaging activities in our own habitat.

With what we know of the geological age of the Earth, how can religiously orientated people come to terms with the concept of Adam and Eve in relating this creation event to what evolutionary theories tell us? If we refer back as far as possible to the first appearance of man-like creatures, in terms of millions of years, then Adam and Eve would be most unattractive and downright ugly beings, and would not seem to be created 'in the image of the Lord'.

The hominids, with their big eye sockets and ridges on the skulls, if existing today would either be in zoos, or in the trees. If the 'big bang' *was* the act of divine creation, and the entire event of 14–20 billion years ago was represented by twenty-four hours on a clock, this would have the big bang occurring at twelve noon on one day and humans appearing at five minutes to twelve noon the following day. Why this great expanse of time before mankind?

The Genesis account, of course, reads as though the whole event was wrapped up, signed, sealed and delivered in a very short time. Poor old Adam had not even got around to all the 'begetting' mentioned in the Old

Testament before the pitfalls and threats began to appear in the form of snakes and tempting-looking 'apples' from the 'tree' of knowledge.

Curiosity is a natural trait of intellect, without which we would learn nothing and must have been *given* by the creator. So the question must be: why give it if He (the Creator) did not wish for Adam to use it? Also, what is so secretive and fearful about this knowledge which was denied to Adam at all costs? Why the punishment for following a natural instinct? In any event, if they had not been cast out to commence the tilling of the Earth, who else would have done it while they roamed leisurely about the Garden eating fruit all day? Once they had begun the process of starting the human race, the Garden would surely have become too small for their needs, and far too over-populated. They would have had to leave and commence all that tilling anyway and of course the clear directive was given to the created beings in Genesis to 'subdue and replenish the earth'. There is still this vast amount of time to consider before the dawn of man. Before we had our dominion over all the other creatures of the Earth in our 'day', many creatures had existed and become extinct before we came along, and so totally escaped our dominion over them. But the creation story seems to indicate we all came along around the same time.

We are told in the Bible that we are made in the likeness of the Creator, but in whose likeness were the dinosaurs created? They appear to have dominated the Earth for something like 180 million years and had disappeared over 60 million years before the dawn of man, so quite obviously they all neatly escaped our dominion, which is probably just as well for them as it is likely they would then not have made it to the period of their mysterious demise.

So here we are, produced in the image of the Creator. Of course, this implies appearance only and nothing to do with our mental condition. Some of us display behaviour patterns that are questionable even to this day, but as well as being capable of savagery we have this fantastic brain, enabling us to achieve as much in the positive mode as we do in any negative actions. Yet a large part of the gross annual budget of so-called civilised nations is allocated for expenditure on improved ways of killing each other, and ever since the first grunting sub-human beat another over the head with a bone we have been busily engaging in wars and the slaughter of our fellow men. Even the fiercest creatures we can imagine in the wild live together in harmony and kill only for food in an open and natural way. Humans make a profession out of killing and do so in a mechanised and mass-produced way by creating armies and weapons for the task. As wonderful an organ as the human brain is when directed towards the more positive and constructive activities, we seem equally at home perpetrating the destructive, wasteful, savage and ecologically damaging actions, and when we are not killing each other on the battlefield, we unite to crucify the planet. This encourages the logical conclusion that the human being is a creation by *fallible* beings rather than a perfect infallible divine creator.

Religious teachings specify that God knows and sees everything, even our most secret thoughts. How utterly boring for Him. How could anyone in that position be envied? Even before He commenced His great creative work, He would know the outcome of it all. He would know all the despots, dictators and brutal rulers that would arise to persecute His people, all the wars to come and who would prevail, all the terrible deeds that we, His creation, would be capable of and would carry out. He would know the end results of every situation. He would even know the ultimate fate of the world, and those who would keep the faith and carry out only good deeds, and all the evil people and what *their* fate would be.

What would be the point of starting it all in the first place? If He simply gave us the choice between good and evil and left it at that and did not know what acts we would commit, that would be a different matter. It would then be all down to us to sink or swim, to be saved or to be damned, and everyone would then be judged by Him before departing to our assigned place for all eternity.

Of course then He would not be all-knowing. He would be as flawed as humans are not a perfect being after all. Few zealously religious people would be able to accept a God who was faulty, as unreliable as us and, perhaps, liable to make mistakes. We must add that if He gave us the choice and didn't know the outcome of our acts, then how could He judge without knowing and seeing our deeds what was good or bad during our lives?

We can read of the actions the Almighty was capable of in the Bible with the pestilences, plagues, floods, the destruction of Sodom and Gomorrah and turning people into pillars of salt just for looking back. It would seem there is a side to Him that is capable of killing without mercy and He becomes capable of being compared to some of His worse creations here on Earth. (These actions seem to represent a 'weeding out' operation to prevent further proliferation of genetic failures.)

However, having gone to all the trouble of creating us, it is logical to assume that He would wish for our continuance, for us to survive and generally prosper, unless He viewed us as a complete failure. But then He would have done something about us long before now. So, since He has had such a huge amount of time observing our behaviour and seen many good and bad deeds and actions perpetrated, He must by now have a pretty good idea of our general behaviour patterns and think that for all our faults we are the best that He (or they) can produce for now.

So is there hope for us after all? Will we escape the Armageddon? But then, He sees the great brain we have been given being used to create horrifying weapons to slaughter each other more efficiently until with one device we can destroy an entire city and everyone in it. Clearly something has gone wrong in the calculations. It simply is not logical that a perfect being could make such mistakes unless He purposely chose to do so. But what would be the purpose of such an act? So what can we conclude *is* in store for us? The year 1999 seems to be significant to the prophets and soothsayers, perhaps simply because it is the end of the millennium. But maybe it will be the 'second coming', the time when all the defective human units will be eradicated?

On the other hand, perhaps we take ourselves too seriously. We may have just been merely an interesting and amusing pastime for the Creator, a pleasant diversion, just something to occupy His time, to fill the vast stretch of time that is eternity.

In spite of the mental abilities we possess and are so proud of, and which are really quite wonderful, the general construction of the remainder of our anatomy leaves a lot to be desired with its obvious vulnerability, and does not lend itself to longevity, affording us little protection from the 'thousand ills that flesh is heir to'.

The only part of our body that is really well protected is the brain (fortunately) with its skull structure, the closest we come to having, like some crustaceans, a skeletal frame on the outside. So it can be seen that there is quite a difference between our skull, particularly its contents, and the body in general – rather like a Cosworth engine in a pedal car. If the poor old dinosaurs could not survive the alleged celestial event that caused their demise with their thick horny skins and better protected forms, what chance have we, made up, it would seem, largely from liquid? Surely a creative God would have given us better protection.

But of course, we have our brain. We have the ability and knowledge to be aware of the dangers, to scan the skies and to observe any approaching cosmic threat. We have the technology to construct devices to protect ourselves from them but we have no such protection. It seems quite incomprehensible that we allow ourselves to be exposed to annihilation when we could prevent it.

Are we to assume that all this dangerous orbiting debris must be all part of the grand master plan of the all-knowing supreme Deity? What perverse, macabre humour to have the odd mountain-sized chunk of rubble hurtling threateningly towards us from time to time and, rather like some vast celestial form of Russian roulette, once in a while hitting us square on, as many a cratered and scarred area now revealed by orbiting satellites give witness to.

Of course, the only other choice is to assume they are mistakes, miscalculations or flaws in the grand plan, once again pointing to a not-quite-perfect Creator who is vulnerable and liable to make mistakes on His grand lofty level just as we are on ours.

The aforementioned seem to be the only two choices available to us. We accept a divinely created universe of originally intended order and perfection, or an unpredictable, somewhat chaotic, dangerous place to exist. We may be thankful we do not exist in the colliding galaxies we can see occurring in deep space, or that our nearest star Proxima Centauri is about to explode and instead of being four and a half light years away is very much closer.

However, it does seem as though we have an 'edge' now, with our intellectual capacity, to be aware of the dangers. It really is quite mind-boggling that petty political posturing, as depicted in the movie *Meteor* for example, should prevent us having some form of planetary protection. There is obviously suspicion of other nations using such orbiting weaponry for possible war-like purposes. With all the great intellects of the world and our current technology, knowledge and awareness, it is still possible that such bodies coming into our atmosphere over a large urban area could destroy all the positive things that mankind has achieved and developed, and all this when we have the ability to prevent it.

It seems to be calculated that major impacts can be expected to occur every 50–60 million years, so clearly we are a lot closer to the next event than we are far from the last. If our Creator does not wish for our demise unless *He* decides it, He will be watching closely to see how we deal with the problem.

Perhaps we need to form some kind of select group, or some other co-operating worldwide think tank, whose ultimate objective would be the total elimination of negative and destructive behaviour patterns and to enhance and accentuate the positive side of our mental processes, neurological surgery or genetic manipulation not being ruled out.

A yob who beats an eighty-year old woman half to death for a few pounds, for instance, must have a certain area in his brain (if he has one) that motivates such action and this could be located and possibly 'adjusted' in some way. It may be the only solution to release us from this 'Jekyll and Hyde' syndrome, with our brain's advancement held back and forever locked in a struggle between civilised and barbaric behaviour patterns. *If* we develop the neurological skills to release the positive side from the restraints imposed by the negative side, surely we would use them?

The human race should by now have the intellectual capability to rise far above all the negative, wasteful, unproductive and uncivilised activities, such as wars, petty squabbles and territorial claims. We should be able to eradicate poverty and work towards eliminating disease, starvation etc. without leaving it all to a comparative handful of dedicated people. It seems Mr Hyde will always frustrate Dr Jekyll in whatever noble aims the latter may have. Human beings were savage when they set out on the road to advancement and retain savage genes even today. Now we have sophisticated weaponry and ways to kill each other more efficiently, so it appears that Mr Hyde is winning the battle simply by holding us back. The Einsteins of the world have proved what we are capable of, but perhaps there are just not enough of them. We may have to wait until we use all our grey matter, assuming that the necessary cells to counteract or nullify negative traits are not yet developed.

Whether we are a serious creation or just an amusement, it will be seen by the Creator that we are in possession of space travel capability; that we could build star ships and set off with a voluntary crew, with recycling systems to the stars; that we are developing a 'star wars' technology; and that, having lost none of our tendency towards violence and war, we will soon be moving into *His* domain. How long will this new situation be tolerated?

Surely now, it would seem, the time has come for the Creator to assess and review the performance and results of His great work, perhaps to think it out again and maybe even to admit He has gone wrong somewhere?

Maybe He will even consider termination, or at the very least drastic modification, in an effort to bring us all back to the original and intended specifications on the blueprints. But perhaps we are not yet considered to be a real threat or danger, although we feel ourselves that we are close on the heels of the secrets of life itself and are moving towards discovering the secrets of the universe: we may be seen as just children playing, or a monkey handling the lock on his cage which he will never learn to pick. It seems that some day we must surely use all our brain material and, if or when that time ever comes, we may be easily able to comprehend eternity for instance, something travelling on forever, without feeling our brain is going to blow a fuse. When we have found the smallest particle and located the end of the universe and are picking freely the fruit of the tree of knowledge, suddenly, the head gardener may come along! But instead of punishing us as we might expect, He may decide that now at last we have come up to the original intended specification, that we are now the 'full shilling', the negative brain cells having been successfully eliminated by the positive ones.

In any event, it is not possible to view ourselves as created beings without imagining the plans, intentions or next moves of the Creator.

One cannot envisage an almighty being coming along that is so capable of such wondrous acts as creating an entire universe, including our sun and planets, then just plonking human beings upon it and wandering off into the sunset, never to be seen again.

The Bible tells us of the final judgement day and Armageddon, and the day of reckoning and so forth. So it seems clear He is going to retain at least a passive interest in us. But how does He find the time to get around all the other billions of Earth-like worlds that must exist in all those galaxies, each presumably having been promised their own 'Day of Reckoning', yet none of us knowing when that may be?

Whether plans exist for us or not, we seem to be very much in charge of our own destiny, and indeed came very close to destroying ourselves during brinkmanship, the Cuban crisis, and the general bluffing and sabre rattlings of the Cold War sixties, with nuclear extinction not only a possibility but a hair's breadth away.

Of course, religious people will say it was the Almighty that intervened to save us. It is certainly worth considering. After all, quite a lot of praying did take place during those times when the missile parts on the Russian ships were steaming towards the American naval blockade.

Of course, all the other religious persuasions would possibly have been praying to *their* gods too, the Arabs to Allah and Orientals to Buddha and so on: it's quite debatable which one did the intervening (if indeed any).

Coming back again to this clearly wondrous organ we have housed in our skulls that we know so little about, the power of prayer or the will or the mind or whatever we call it is no doubt a force to be reckoned with. It seems we can 'will' ourselves to be unhealthy, just as a positive outlook on health and our well-being seems to make us feel better.

Stigmata, the appearance of the wounds of Christ on the human body, seem now to be accepted as caused by the minds of over-zealous religious people, just as cures at Lourdes seem also to be a product of a strong will to 'believe', and therefore to be cured. Voodoo priests are well acquainted with the mind's ability to not only cure us or make us ill, but to die at the appropriate moment specified without wasting anyone's time. Of course, the essential ingredient is that the victim must *know*. Then, of course, his or her mind takes over.

Religious people are perfectly entitled to their beliefs with regard to the power of prayer and Lourdes etc. as being divine acts, and it would be wonderful if they are right. While my views seem clinical, detached or atheistic, I would just add that I hope I am wrong and they are right.

Personally, I think we are only just scratching the surface in our discoveries of what the mind is capable of. Once we separate the stage-managed tricks of the illusionists from reality, there always seems to be a handful of people from various countries in the world who can perform that which the rest of us have an inner feeling we could also do if we had enough power of concentrated thought.

In the Bible God said, 'What I can do so ye also can do'. One wonders if we have started on the road to this culmination of harnessing the full power of the brain. Of course, the old problem of the Jekyll and Hyde syndrome now rears its ugly head. If we haven't eliminated it by the time we do harness all these forces and fully use our maximum brain potential, then God help us. We had better make sure there is an equal advance in eliminating these negative qualities as there is in achieving mind potential.

It is quite incredible to reflect that, since the onset of technology, which one could suppose began with the mining of metals, the greater part of our energy and efforts and annual budgets has been the manufacture and development of more and more sophisticated weaponry and general hardware for war-like pursuits. There is obviously something very wrong somewhere.

What of the Creator Himself? Who created Him? Of course religious teachings neatly side-step this question by stating, 'He always was and always will be'. To tell someone to have faith and believe is simply another way of saying, 'Put your mind at rest, don't think, just believe and you will feel better'.

One could equally state black was white and, when questioned on it, just reply, 'Trust me. Just believe me. Have faith in what I tell you.'

It is perfectly possible that creation is, in itself, just *that* – a 'creation' concocted by the brain of man as a direct result of his own self-awareness and feelings of importance in the scheme of things. He therefore assumes that all other creatures were created simply for his benefit and enjoyment, neatly ensuring the clause of his dominion over them is included in the write-up as part of the deed, covenant, policy, Bible, or whatever we want to call it. He assumes that the birds were created to give him pleasure with their songs, chirpings and graceful wheeling about the air above him. Their other purpose, as well as this, is to eat up all the flies and bugs that pester him.

Then there are the fish of the seas and rivers, created to provide him with a different source of food. The cows, pigs and sheep of the field to provide his main diet and to cover the bodies of himself and his family with skins to protect them from the winter cold.

We can carry on quite happily in this way if creation is an invention of our own mind, but if we do believe in divine creation, who gave us leave to devour other creatures who are themselves created beings and share this world with us?

Imagine our shock and horror if we found out that the scenarios depicted in science fiction literature are true and we ourselves are just bred as 'food for the gods', and that the countless people who disappear every year wind up on some extraterrestrial dining table? Why not? Are *we* not doing just this? Do we not breed creatures simply to eat them?

Surely all these other created beings have as much right to their lives as we do. But wait a minute – what about the slugs, the curse of the proud gardener? Even the birds won't touch them. The snails leave their slimy trails everywhere, the wasps (unlike the bees) only give us a sting. Then we have the deadly snakes, spiders and scorpions, lions, tigers, bears – creatures that would terminate man's life without a second thought. What was the purpose of *their* creation? Was it all part of the same macabre sense of humour of the Creator to produce hazards, pitfalls and stumbling blocks, dangerous threats to our existence, just as the celestial comets, asteroids and meteorites were introduced?

If we believe in a divine Creator and that the biblical version of the event is essentially correct, then surely we must accept all the other religious teachings, doctrines and concepts also – the hereafter of Heaven and Hell, eternal damnation or a continued spiritual existence.

Again, this eternal struggle with positive and negative, good and evil in opposition to one another, comes into the picture.

Given, as previously suggested, that Heaven must be a very crowded place if beings go there from the countless other worlds that surely must exist, so Hell must be equally as crowded. No wonder demons are reputed to say, 'We are legion' – they would be indeed.

If we assume that Heaven and Hell are only the prerogative of the Earthly deceased, then it comes back to man's arrogance in assuming that his is the only world that is populated and important in the scheme of things and that everything was created with him and his world in mind.

In spite of the darker deeds our brains are capable of, it is an orderly instrument and craves organisation, neatness and order, with stability, out of preference to anarchy and chaos. It is this ability that allows us imaginatively to see canals and faces on Mars, structures on the moon and, perhaps, mirages in the desert. And, to be sure, if there had not been a biblical description of the event on Mount Sinai, then we would have invented our own legends. We need an ethical code or series of commandments to live by or there would be chaos, anarchy and misrule. Mankind's brain tells him that a code of behaviour is necessary to his existence and well-being on the planet.

It is perfectly possible that the biblical version of the issuing of the commandments is a concoction of the brain of man, couched in dramatic terms with unseen, booming voices, thunder, lightning and other things to heighten the drama.

We only have the writer's word for it ñ no actual stones to analyse and no Ark of the Covenant to study (in spite of the efforts of Hollywood actors).

Humans are thinking, reasoning creatures; we do not act on instinct alone. (As a matter of opinion, I do not believe domestic animals such as dogs and cats act on instinct alone either, as they can be observed to dream, twitch, run in their sleep etc. One cannot dream without the ability to think, for what is a dream but a series of events running through the mind that one *thinks* is happening.)

But man, without rules, laws and a distinct code to follow, would become degenerate, degraded and sub-human (as indeed it would appear some of us are capable of being, even with a code).

If it is true that Heaven and Hell are a product of the inventive capabilities of the brain of man, then it becomes even more clear that the mind is trapped in this situation of civilisation and savagery being locked up in the brain in equal proportions. Clearly, Heaven and Hell now become a mental trick to enable the positive side of the brain to overcome the dark side by making us sit up and think with frightening mental pictures of the torments of Hell and its horrors, to gain an advantage and direct the brain patterns to the positive side. So here we are, back to the same old conflict of good and evil.

If the ultimate aim of it all is to produce as many good spiritual beings as possible to fill the portals of Heaven, why were we not all created perfectly good spiritual beings in the first place?

Then we could go straight there without all the trauma, temptations and pitfalls put in our way that, if we succumb to, will send us straight to the other place. The teachings specify that we are put on Earth with this choice between good and evil by an Almighty and All-knowing Creator, who must be already aware of those who will make it and those will not in the first place. It all seems such a waste of time unless the whole thing, as said before, is simply a grand production of entertainment for the Creator and, even though He knows the outcome of it all, He will have great fun observing it.

If our advancements continue to streak ahead so fast in other positive achievements, we will have to do *something* about the retarding, negative factors in the brain continually holding us back. If we assume we *are* divinely created beings, who do we blame for our restrictive mental condition that has produced the Caligulas and the Hitlers? And we must also ask, what are His intentions about alleviating the problem? How will He rectify His mistakes? Do we believe in stories in the Bible that previous attempts *have* been made with the divine holocausts sent upon us as the punitive actions of the Creator, where thousands perish and entire cities are laid waste, all down to faulty behaviour patterns in His creations? Vast land masses, even entire continents were drowned in massive flooding, where only a handful of correctly functioning beings are allowed to be saved in their Ark, along with all the different types of animals? (The Ark, of course, was constructed by a specific guidance from the Creator.)

Was this a brutal and crude form of genetic cleansing, where only the good and more positive thinking beings were allowed to flourish and survive and, hopefully, beget more like-minded offspring?

Was the Nazi holocaust, and is the ethnic cleansing so recently heard of in the former Yugoslavia, simply the result of inherited racial memories, deeply planted in the brain stems of beings created in the likeness and image of the Creator and carrying out like-minded actions in order to eliminate lesser mortals not considered quite up to the mark?

Well, clearly, the genetic cleansing by the Creator failed also, for here we are with the same old failings letting us down, earmarking us for the pit as not quite right and not up to full specification. *Are* we going to be drowned, burnt, turned to pillars of salt, or have a pestilence descend upon our house and perhaps a plague or two?

What is the plan? Surely the mass murderers, the drug suppliers that cause such misery and death and are responsible also for the ravages wreaked on innocent people's homes as their victims search for money to feed their sorry habit in their pathetic struggle towards their next fix, the perpetrators of organised crime, the grinning third world dictators and sub-humans who purposely starve their populations in order to buy weapons cannot all escape 'rectification' and adjustment forever?

Surely, all this mayhem going on all over the world and all around us that provides grist for the mill of the media that we are also becoming dangerously used to, will not be allowed to continue forever unchecked? The beasts of the field quietly continue to chew the grass, oblivious to it all and patiently awaiting their turn to be eaten in blissful ignorance. The birds continue to sing to us. Even animals who are free from captivity in the wild only kill for survival and take what they need and lie down and lick their cubs. So what is the Creator going to do about it all? Is He sitting on His hands, His mind wracked with indecision, still pondering what to do, wondering where it all went wrong?

Or is He just coolly surveying us and patiently watching for some signs of gradually improving behaviour up to the deadline He has set (which some prophets of doom seem to specify as 1999, or the end of the current millennium). To be sure, if we are indeed created beings, then clearly the Creator is surely neglecting His creations. We are certainly in need of *some* adjustment and fault diagnosis, perhaps a 2,000 year servicing and correction. The question is, when will it happen? *Everyone of us* breaks at least one of the ten commandments *every day*. The ten commandments could have been written by *any* wise ruler, king, potentate, Pharaoh or leader or even by *Moses himself*. Moses had the advantage of an upbringing in Egypt at least equal to that of the Pharaoh with regard to his education and tuition and he clearly saw the need for a behavioural code in the Hebrews who quickly reverted to worshipping graven images the minute his back was turned during the desert wanderings after the Exodus. Certainly no one would deny the need for a code or list of human directives. Seeing the *need* for them is the easy part; *obeying* them all seems to be the hard part for many of us.

It seems strange that the Babylonians and other great empires such as the Egyptian dynasties that preceded the Exodus arose without feeling the need for such a code, charter or constitution to guide their people. Perhaps if they had have thought of producing such guidance, the iniquity of cities like Babylon, Sodom and Gomorrah may never have existed.

Stones, much older than those brought down by Moses from Mount Sinai, exist making up the grand edifices in Egypt. Gold, a noble metal, lasts almost indefinitely. Therefore both the tablets of stone and the golden Ark of the Covenant in which they where placed could *still exist*, unless smashed up or melted down by the enemies of the Hebrews.

With regard to human behaviour and our quest for 'civilisation', whether we regard the commandments as 'sacred' or otherwise, one fact which *must* remain sacred and must be learned by all humanity in every land on every part of the globe if we are ever to aspire to becoming *truly* civilised is the recognition of the value of *a single human life*.

2. Close Encounters in the Bible (The Angels of Abraham)

The alleged close encounters of the third kind with possible unearthly beings that supposedly take place today are either abductions which can be brought out from the mind by hypnotic regression sessions, or people accidentally coming upon an unidentified landed vehicle. In both cases there is a reluctance for conscious close contact with humans. Some sort of mind scan, to calm the victims and make them forget the incident, occurs in the case of abductions. In other cases the beings quickly depart in their craft, exercising speed and manoeuvres beyond the capabilities of earthly craft. In the days of the patriarchs, certainly in the case of Abraham almost four thousand years ago, beings moved freely among them that the Bible refers to as 'angels'. There are many people who would offer quite a different interpretation of 'whirlwinds' coming down from the sky and then departing taking biblical characters with them, of 'wheels' with eyes all around them, of creatures with suits of burnished bronze and whirring wings, chariots of fire, flying shields, and so forth.

The list of events in the Bible that appear to have extraterrestrial connotations is enormous. Elisha witnessed Elijah being taken up to Heaven in a 'whirlwind'. What's more, Elijah knew he was going and discussed the event with Elisha beforehand. These beings, whether we accept them as angels or aliens, had a direct concern, and became very involved, in the affairs of mankind. They carried out very severe programmes of a type of genetic cleansing or 'culling' of what they saw as undesirable human rejects only fit to be terminated. There was more celestial activity on the night of Abraham's birth than that of Jesus with *his* moving star. Hebrew legend states that King Nimrod's magicians spoke of a bright moving star in the east 'swallowing' *four other stars*. Here we have a clear image of a mother ship taking its small planetary observation craft on board. Why are the angels and notables of the biblical times shown with what appears to be a dome or sphere over their heads which we call a halo? Although angels are depicted in this manner, clearly the angels of Abraham were a far cry from our usual pictorial image of them on Christmas cards. Were those 'halos' protective breathing domes?

The patriarchs themselves seemed also to have special powers and seemingly worked their share of 'miracles'. Could it be that they (and this may include Jesus himself) were a kind of hybrid being born of human females, but with extraterrestrial genetic material in their makeup, with those aerial craft seemingly controlling or supervising their birth process? Abraham also seemed to have the power to resist being burned at the stake, when the wood did not ignite and the priests swore that an 'angel' was responsible for this protection.

Can we read 'alien' instead of 'angel'? Clearly, they were very humanoid or could take on the human form in a convincing manner in order to move among the patriarchs. From a modern day viewpoint, the only way for us to interpret a moving star would be a meteor, which would quickly disappear, or an orbiting satellite, which would hold a steady course until it disappeared over the horizon. Any 'stars' that move about or travel slowly enough for three men to follow them on camels, then stop over a town or lead people through the desert at night with a beam of light, cannot be interpreted as anything other than a controlled aerial craft.

The 'angels' moving so freely among humans in those times would know that they would receive adulation and reverence by technically unadvanced and unsophisticated humans of those times, just as in old days of human exploration natives fell down and worshipped people from advanced cultures coming along them. Were these 'angels' of biblical times the *descendants* of the original genetic creators of mankind? Why would they wipe out thousands with floods, pestilences and nuclear explosions if they did not see their 'victims' as genetic failures of freaks? These moving stars, if they were controlled aerial vehicles, were clearly 'not of this Earth', the very words Jesus used when describing the location of his kingdom. Was the moving star that supervised the birth of Jesus the same craft he 'rose up' to after revival by the 'angels' who stunned guards (with what – 'phasers' of some description?) after claiming his body from the tomb and restoring his life functions.

Although the patriarchs, such as Abraham and Moses, had access to the mountain tops from where the booming and roaring and pillars of fire and commanding voices came, the *masses did not*, and the 'angels' feared their curiosity. Moses was told, 'Go down, charge the people, lest they break through unto the Lord to gaze and many of them perish'. Was this a death threat for just looking, or was their power source a real danger to indigenous earthlings, with its dangerous protection shield?

The 'angels' never displayed their powerful weaponry or means of delivery, but simply ensured that any particular beings whom they had chosen to survive were led from the area of destruction.

The 'angels' ate and drank with Abraham, and in the case of Lot and his family as they 'sat in the gate' of Sodom, they feasted, drank and stayed overnight as guests in their house on the evening before destroying the city.

That evening, the decadent and depraved men of the town came to Lot's house and called for them to come out. They wanted to 'know' the angels, which in biblical parlance meant they wished to have sex with them. For this impertinent request they were blinded. (With what? The same 'Star Trek' type phasers?)

The next day, Lot and his family hurried away so that the angels could use the weapon. 'Hurry to the mountains. Look not behind thee, for I cannot do anything until thou become thither.'

Was a similar weapon used on another iniquitous city, Babylon, situated in the Euphrates Valley which is now southern Iraq? It would appear that an archaeological dig went down through various layers, then stopped at a hard, glass-like, vitrified layer similar to Alamogordo nuclear test sites in Mexico. These 'angels' certainly metered out some pretty severe genetic cleansing methods in ancient times with plagues, pestilences, floodings and what appear to be nuclear weapons: the 'angels' of Exodus assisted Moses to drown an entire army and 'went forth' before the Hebrews and gave them instructions: 'Let not any creature that breathes, to live'.

If we choose not to accept these 'angels' as divine beings (which they may very well be), we would have to conclude that the writings are all a complete invention or that the 'angels' are some form of extraterrestrial beings with enough interest in mankind to have possibly been responsible for our creation, and thus felt quite obliged to terminate humans not manifesting the right characteristics at a stroke. What is their point of origin? For want of a better description, could we call it 'Planet Heaven'? Was the controlled aerial craft that led the Hebrews in Exodus going forth as a 'pillar of fire' acting on instructions from Planet Heaven?

Was the same direction of mankind by the *descendants* of the 'angels' still underway some 2,000 years later with the emissary Jesus?

3. The 'Emissary' Jesus

Can we conclude that the ruling power on Planet Heaven had moved away from away from its hard line policies with regard to the human question? What a contrast with Jesus and *his* policy doctrines of humility, love they neighbour and turn the other cheek, compared to the plagues, pestilences, floods and bombs of earlier times, with people who were supposed to be righteous being turned to pillars of salt for *just looking*. Obviously, the ruling body of any far-off alien world *would* direct the actions and methods of its cosmic voyages on other worlds, and this complete reversal of policy implies a certain impatience, uncertainty and unsureness about just how to deal with their early 'creations', and the 'human question'.

One cannot fault the teachings of Jesus and, even if we do choose to interpret Him as the equivalent of modern day Hyde Park evangelist with a gift for convincing magical tricks *or* an alien emissary, his teachings were no less valid for all that. What did he preach that warranted such barbaric termination? During his sermons he was careful to keep outside the political arena and should not have been seen as a threat to any particular ruler. To be sure, it did not help his case when people chanted, 'King of the Jews'. But his *main* mission was clearly to strongly influence human behaviour patterns. Even though the 'angels' revived him, there must have been enormous debate on 'Planet Heaven' about yet another example of human barbarism and possibly even a suggestion to terminate the plan, destroy all humanity and perhaps start again from scratch.

If the ongoing UFO phenomena in our skies is the continual overseeing and progress analysis of their creation's *descendants* (i.e. *us*), perhaps we had better hope that the hard-liners do not appear again on 'Planet Heaven', or we may begin to see places like Soho, the Reeperbahn in Hamburg, seedy areas in Bangkok and other vice areas of the world, disappear in a cloud of smoke. Perhaps 'Plan Jesus' was even seen as a 70–80 per cent success, as humans were actually observed to forfeit their lives by being torn apart by wild beasts rather than renounce his teachings, and others had travelled far and wide to preach the message. Human behaviour patterns must have been seen to change from the days of Babylon, Sodom and Gomorrah, even though wars and negative behaviour still continued.

However, if alien beings *have* been attempting to direct human behaviour patterns for as long as it seems to indicate in the biblical references, then we must consider the countless wars, slaughters, enslavements and domination of other lands, right down to the holocaust in Europe, Japanese atrocities and nuclear weapons actually used in anger, that 'they' have witnessed over their lengthy observations.

One could easily imagine a great complex or memory bank facility existing in the future in somewhere like the NASA organisation, keeping track of earthly cosmic voyagers and their activities on far-off worlds, storing information on life form studies and experiments. It is easy to imagine the situation in reverse, with beings, perhaps thousands of years ahead of us, looking at *us* as their experiments. It can only be concluded that if the UFOs are a manifestation of this intelligence in an ongoing programme, they have *enormous* patience with us, given the questionable human behaviour patterns on the negative side over the last few millennia. But, if we accept the possibility of their existence, we must accept that there may be a conclusion to the overall plan. Will it culminate at the end of the current millennium as so many prophets and soothsayers have hinted? When will 'they' consider us ready for such fantastic revelations? After all, we are making serious plans for deep incursions into *their* domain. *Will* 'the great king come from the sky in 1999 and seven months', as Nostradamus appears to have foreseen?

4. Eden: Human Creation Zone?

The Bible specifies that there *was* a creation centre for mankind, but who were the group, the board or committee that made the famous statement, 'Let *us* make men in *our* image'? Were they an assembly of highly skilled geneticists? If the angels of Abraham were so human-like as to mingle with the patriarchs and carry out normal human functions like eating and drinking, it would appear that we *are* made in their image.

They may also be responsible for the three distinct types, Caucasian, Negroid and Asian that make up the human race. With our sparse collection of bone fossils, we are forced to construct models and produce drawings to support human evolutionary theories. *Which* of the various types would result if hypothetical alien beings chose a simian creature to circumvent many millions of years of natural, but excruciatingly slow, evolution?

The hominids, who would look more like gorillas, *Homo erectus*, with a supposedly upright posture but with a skull like an ape? Neanderthal man, allegedly squat, beetle-browed and ugly? Or Cro-Magnon man, with a highly developed brain, upright and intelligent? Clearly, it would be the latter. If so, did this genetic experiment begin shortly before the advent of Cro-Magnon man, some 35–50,000 years ago?

The Euphrates Valley has been suggested as a likely area for this mythical garden. Even to be discussing a likely place implies a comparatively recent timescale for the event, and not in terms of millions of years.

When Abraham was born in Ur in Mesopotamia and began to prepare for the actions he may have been '*programmed*' to fulfil the population distribution plan eventually undertaken by Moses. Egypt was *already* highly advanced and settled. The most intelligent and enlightened people on Earth on those days would have been the Egyptian priests. Is it beyond all reason, or can we look at the possibility *that the lost island of Atlantis* may have been the creation zone of mankind? The Atlantis legends stemmed from the *birthplace of Moses* via the Egyptian priests to Solon the Greek and to Plato the writer.

History regards the ancient Greeks as being highly refined, advanced and producing great intellects well ahead of their time and, indeed, all that *is* true. However, the Egyptian priests looked on them as naïve children, bright, but still learning. The Atlantis myth did not originate from the literary works of Plato, but from the Egyptian priests, who hinted that they had records of enormous periods of Earthly history to call on. Plato put its location firmly in the Atlantic. (To assume Thera in the Aegean was Atlantis is to assume Plato was an idiot)

The catastrophe that befell this island between the Americas and Europe, was said to be only one of *many* such occurrences that they knew of and the most profound revelation made to visiting Greeks such as Herodotus and Solon, was an apparent awareness of the 'precession of the Equinoxes', or slow movement of the axial inclination of the Earth. They stated that the sun had not always risen where it did then, indicating a previously different inclination of the Earth's axis, or a *period of history of 26,000 years of polar gyrations* (rather like a spinning top slowing down). This length of time almost reaches back toward the appearance of Cro-Magnon man. This mythical island supposedly collapsed into the sea around 10,000 BC. If humanity *was* created there in some alien Garden of Eden, the descendants of the first creations could indeed have built up a fantastic culture, watched over by the amused and rather proud alien descendants of the original team that had carried out the original plan of 'making men' in their image.

Do the mysterious origins of the Egyptian culture that *did* seem to come into being rather quickly have *their* beginnings on this lost island that the priests seemed quite convinced *did* actually exist? Were humans created *only* to till the Earth with the sweat of their brow, or to become ultimately cosmic creators themselves?

Or was it perhaps both? A safe island habitat *would* be a most suitable location. The mid-Atlantic Ridge may have been less geologically active 50,000 years ago, and an island would contain their creations until they were seen as being fit to be transported. *Are* all the creation legends simply racial memories of these distant events, particularly the North and South American Indians? Everyone else assumes that their origins are in Asia and that they trekked across the Bering Straits into the Americas. *They* say they originated to the east, or the Land of Morning, and the North American Indians say all the tribes once lived together *on a large island.*

In any event, the possible alien creation programme was well underway and progressing when the alleged loss of the island occurred. The created human intellects may have built up the fantastic culture that Edgar Cayce spoke of when in a 'trance' condition, and the Egyptian culture was a pale attempt, along with South American ones, to revive it. Hypothetical alien 'creators' observing all this dangerous earthly geological instability may have sought a much less precarious habitat.

Perhaps the aliens' descendants moved to the mountain tops after the loss and gave out their commands to their chosen beings, and dropped the manna, or alien food source, to the multitudes of stateless people in an ongoing Earth zone-population programme, and led them with 'pillars of fire' to their destinations. Was the trauma and destruction of early humanity in 'the flood' that found its way into subsequent early legends actually this island creation zone going *down*, making it appear like the waters rising?

How do we sum up the long-term alien plan? We are seriously discussing the terraforming of other worlds and becoming cosmic creators ourselves. Was it:

1. The initial discovery of a planet of the apes growing wild followed by the production of genetically created beings.
2. The tilling and nurturing of the Earth, the 'first lessons' around the time of the first civilisations, ongoing guidance.
3. 'Culling and eliminating', the patriarchs, and final emissary Jesus.

4. The abductions for genetic progress and analysis, followed by the 'second coming', revelations of our origins?

Is the most obvious clue to all the above the human brain, particularly the forebrain that gives rise to the human qualities? Was the temptation in the Garden of Eden simply another racial memory of a alien experiment, or a test for the strength of character in their creation? They may have seen *failure* in the fact that Adam clearly failed the test, but *success* in that he obviously desired more knowledge by removing an apple (information sphere?), supposedly left within his reach from the great information bank or 'tree of knowledge', which Genesis states was specifically positioned in the *centre* of the garden.

Did they then lose patience with him and ensure his transportation to commence Earth's agricultural programme? Do the *descendants* of the alien creators view their biggest failure as the lack of any *further* surgical and genetic enhancing operations in their creations having been carried out after their hard-line predecessors had finished their brutal 'genetic cleansing' programmes, for it seems to have failed, as the 'dark side' of humanity still exists.

Even though the best of the creations have been 'cut out' from the herd, so to speak, the human race *still* goes off to war and slaughter each other from time to time, and has done since any assumed 'alien creation' programme began. So it would appear that all the biological methods of destruction together with the atomising of cities and drowning of thousands has been waste of time. However, we may ask, what would the human race now be like (if it existed at all) if such people that made up the populations of ancient Babylon, Sodom and Gomorrah *had* been left alone to further their kind? However good, bad or indifferent, here we are, either a product of all the above or divinely created. Earthly human evolution seems unlikely for it never would have made it if the end of the line was 'Neanderthal man', as he seemed to be retrogressing and not progressing at all. Cro-Magnon man was not related to him, so it would appear we are either created in God's image or that of the extraterrestrials.

5. Where was Eden?

The following is an extract from my recently completed work, *The Angels of Abraham*:

> We are able to determine a very rough location for the Garden of Eden but the first book of Moses, called Genesis, seems to have been written with the intention of being purposely vague with regard to the precise location of Eden. Genesis states: 'and a river went out of Eden to water the garden and from then it was parted and became into four heads, Euphrates, Pison, Gihon and Hidekel'. Strangely, Gihon is said to encompass the whole land of Ethiopia. Now, we are in East Africa. With East Africa coming into the picture we have an enormous tract of land to consider that stretches from the Euphrates, (the only river we can identify) in a south westerly direction across the Arabian Desert, then finally across the Red Sea into Ethiopia.

Our search for Eden is now likened to that of Eldorado, Shambala, Camelot, the Ark of the Covenant, or the Holy

Grail. *The Holy Bible*, a mature version printed in 1955 by The British and Foreign Bible Society and used for my research when compiling *The Angels of Abraham*, clearly attributes the writing of Genesis and human creation to Moses. Since Moses and his preceding patriarch Abraham both travelled extensively in these zones, it is quite natural that Moses would assign the location of Eden to the land he was familiar with, but if one considers the extraterrestrial hypothesis for human creation, then it would seem unlikely that such beings would consider such a wide open, barren and hostile region for their experimental life form creation zone. Surely an island habitat with the protection it would afford, especially if cleared of predatory creatures, *would* be a better choice. Their creations could then be contained before being dispatched to 'subdue' and replenish the Earth.

Volumes have been written on the possible existence of the mythical island of Atlantis. Plato's imaginative writings of the activities on the island, such as the commerce, trade, the ports and elaborate entrances and waterways clearly reflect the sea trade, organisation and activities of his own time. Quite a different level of technology was indicated by the later Edgar Cayce with his descriptions of the advanced island technology. There is a considerable amount of circumstantial evidence for the existence in reality of this large island land mass, and its possible location being just where Plato said it was. A case for its reality was well presented in *The Mystery of Atlantis* by Charles Berlitz (Panther).

Consider the case that over 50,000 years ago, the Canary Islands and the Azores, instead of being Atlantic islands, were the mountains of Atlantis, and that to our hypothetical alien beings arriving on Earth, this venue seemed ideal for their base to begin their Earthly life form study programme. Their mission may have been to enhance creativity and intelligence in any creatures with an observable tendency toward developing the seemingly ideal humanoid shape. At that time, Earth would have been a 'planet of the apes', and growing wild and uncultivated. Could we assume, after choosing a suitable being, perhaps Neanderthal man, that their plan was to produce a more intelligent being and allow it to proliferate on this large island, then, ultimately, transport them around the globe to till and nurture the Earth, with the sweat of the brow? Eskimo legend states that their ancestors were brought there on great 'birds' (aircraft?) with 'brazen' (metal) wings.

Firstly, they would clear the island of the most seriously predatory creatures, but probably not *all*. It would be necessary for their creations to be aware of all Earthly life forms and to learn to defend themselves from them all when necessary. An island location would make the supervision, location and control of the 'creations' much easier by preventing them wandering away to other lands, and to be on hand for analysis and progression checks on their development.

If alien intelligence did 'produce' Cro-Magnon man by genetic experimentation on this mythical island, clearly it was thousands of years before its assumed loss around 10,000 BC. The aforementioned chap named Edgar Cayce, who was a psychic researcher and ESP investigator, spoke lucidly of a great technology existing there when in a trance-like state. he spoke of advanced flying craft, submarines, electronics and great power sources. Is there a link with the vimanas and weaponry spoken of in old Sanskrit legends? It would not be alien technology that was spoken of by Edgar Cayce, but the result of the efforts and directions of the more intelligent creations from the original experiments, as they naturally progressed and advanced there for a very long time. The ancient Egyptian King Mena (Menes) was said to be an ancient 'flood' survivor before founding the fantastic Egyptian culture.[1]

A link is often made with the fantastic Egyptian era having its roots in Atlantean mythology. Are the ancient Egyptian edifices, as wondrous as they are, merely pale shadows of what may have been evident on this Atlantic island? What are the huge Cyclopean stone constructions found under water, located near the Bimini islands in the Bahamas that were discovered and bore out Edgar Cayce's prediction of their discovery in 1969, by underwater explorers?

Clearly, the object of any genetic experiment to produce intelligent beings would be to obtain the best results possible, although today, genius seems to be very well spread out in the great gene pool, or morass, of Earthly populations; occasionally geniuses, child prodigies and great intellects *do* surface from time to time. Imagine a whole race of them existing on this mythical island. But now, except for the clearly advanced intellects that arrived as survivors on the South American coast and the Mediterranean areas to teach and initiate and possibly attempt to recreate their lost technology, they are all drowned, along with the destruction of this wondrous technology.

The Egyptian culture did seem to appear rather quickly without any specific history or roots and a lot of mystery surrounds the various South American cultures and their origins such as the Maya and the Inca civilisations and

---

[1] Rene Noorbergen, *Secrets of the Lost Races,* Nel, 1980.

similar edifices and mysterious constructions spread out all over South America, some of which may be completely overgrown in the jungle areas.

Was this mythical island the birthplace of human genetic creation? Looking at its assumed location, straddling the volcanically active mid-Atlantic Ridge, we might assume it to be a bad choice for setting up an experimental complex. That may be true today, but was the fissure running down the Atlantic caused by very severe geological stresses experienced by the *entire Earth* some 10–12,000 years ago? The event that caused all the floods, the melting of the last ice age, the deaths of so may animals, sweeping them along to their final resting places in great jumbled heaps, the flip of the Earth's magnetic field and other mayhem and upheaval? Not even an advanced alien technology could halt or prevent such a traumatic event as the *possible shifting of the angle of tilt in the Earth's axis,* or stop such volcanic effects caused by plate tectonics.

If alien beings did create mankind in such a location, they *would* transport them out to various locations around the globe in a kind of population distribution programme, as clearly all areas of the Earth would be *swamped* with vegetation. Certain beings or 'creations' showing the most organisation and leadership qualities, would logically be chosen to lead each group. The primary object of such an operation on Earth would be to produce intelligent beings to nurture the Earth and to make the best possible use of such a (probable) rarity in the cosmos.

It is quite debatable whether unearthly beings, which may be the descendants of the assumed creators, regard us as properly fulfilling the role today when observing our ecologically damaging actions. During these fantastic genetic experiments, although seeing curiosity as an attribute of intelligence in their 'creations', there would be certain things that would be off limits, so to speak, that they would not wish to impart to them or have them enquire into. Any especially intelligent beings who attempted to do so would be transported as far away as possible to proceed with the Earth cultivation programme, which was the original intention of the creators for their subjects in the first place.

As said earlier, in spite of the assumption that the American Indian races trekked across the Bering Straits from an Asian origin to populate the Americas, their legends, both of North and South America, have their ancestors coming from the *east* from a large island or 'the Land of Morning', or stating that all the tribes once lived together on a large island.

Once the great process of creating intelligent beings was underway, it would be self-generating and ongoing and they *would* 'go forth and multiply'. The prerequisites for proliferation and advancement were now firmly lodged in the neurological cellular structure within the skull and at the appropriate time would manifest themselves, and once the created 'hybrid' beings had all been transported out and *were* multiplying, there would be no further need for the 'creators' to remain on Earth. They would then depart into the cosmos, but their actions and operations would be carefully inserted into the memory banks on their home planet and further visits to Earth would obviously be necessary from time to time.

If such a creation centre did exist, the activity there *would* quite probably be far in advance of the lands that the created beings, that is our Cro-Magnon ancestors, had been transported to. One could use the comparison between the activities of NASA and what is going on in the Amazonian interior. However, this would have been of little consequence to the hypothetical creators. After all, the primary purpose of their creations was for them to use the Earth wisely, to hunt, clothe and care for their families, plant, cultivate, nurture and receive the bounties of the Earth and put back much of what they receive, and so-called primitive societies *did* carry on these pursuits until 'advanced technology' caught up with them.

Old ways started to disappear and they were introduced to diseases that they were unfamiliar with: crime, vice, slavery and exploitation. They were on their way to becoming 'civilised'. Perhaps this is the price to be paid for technological advancement. We could never envisage a law stating, 'As from tomorrow there will be no more progress'. But who has paid the price? Have happy hunting grounds become perhaps, underground or surface nuclear test sites? Eventually of course, all so-called primitive societies will 'catch up' and the Earth's races will advance together.

This fantastic mythical island may have been intended as the nerve centre of administration in order to supervise the created beings' efforts to nurture the planet, but the creators would surely know that with such intellect, their creations would not be content with such basic and mundane lifestyles, especially with such curiosity, creativity, sense of destiny and intelligence bequeathed to them from their creators. In spite of all the mayhem on the road to advancement, would not the creators themselves have gone through such painful processes long ago on the road to *their* advancement? Perhaps this explains the enormous patience that would be required of beings that have seemingly been overseeing human activities for so long.

Finally, there comes this cataclysmic loss and destruction far beyond the control of, and perhaps totally unobserved by, the (now departed) descendants, of the original 'creators'. However, in any disaster there are always some beings that would survive before dying out naturally or departing (who knows where) and promising to return someday. They would wish to impart as much of their wisdom as was possible to their hosts in the lands they arrived in. *Did* the extraterrestrials' descendants come back for the survivors and take them to their own world *after* these 'initiators' had imparted all this wisdom to other lands and the 'creations' therein? The initiators may have included Quetzalcoatl of the South American legends and Oannes the Babylonian 'teacher'. Clearly, this knowledge was well utilised and the flowering of civilisation began. Priests or chosen ones would have been carefully selected as keepers of profound knowledge. Is it possible that something along these lines actually happened? There would be some attempts to regain some of the grandeur that may have existed on this mythical island, and this may have manifested itself in the ancient Egyptian culture. How much knowledge *was* entrusted to the wise ones or priests? How profound was it? Will we ever discover it? Coptic texts speak of King Saurid, instructing the priests to bury secret knowledge in a pyramid; is that what the scientific programmes were looking for instead of kingly burial chambers when carrying out the X-ray programme using natural cosmic rays passing through the pyramids, and special recording equipment in which the computer printouts would show darker areas?

What about the tomb robbers? Was this profound knowledge found and simply destroyed along with all the other scrolls and knowledge, to heat the grand palace of some later conqueror? Perhaps the priests were a bit wiser than to bury it where it could be too easily found. Is it simply two blocks down from the pinnacle? Once the pyramid was completed with its smooth cladding in place, no one could scale them. It may not necessary at all, or even desired by an extraterrestrial intelligence now possibly in Earth space, to appear to mankind in order to enlighten us about our possible beginnings. They may only need to direct us to the source of this possible hidden knowledge. The pyramids are 'scaleable' now the cladding has been stripped away. What would an x-ray of the upper regions reveal?

The pyramids seem to exude an air of mystery, and it seems hard to assign their construction to human endeavour alone, or that their massive construction effort was simply for a burial chamber. However, there still remains the mystifying enigma of the highly endowed human brain, defying normal evolutionary processes, and it is the most profound piece of circumstantial evidence for some outside force or influence being responsible for it, and if a member of the simian kind was chosen by unearthly beings to produce an intelligent being to regulate the Earth, then those much quoted genetic similarities in apes *would* be apparent. In any case, as said in the introduction, pigs have contributed their genetic material quite successfully to humans in certain heart operations. All creatures have genes, DNA and chromosomes and this may apply to all creatures of the universe, let alone of the Earth. If the list of elements is finite, we will all be made of stellar material, that is, atoms of matter once blasted out from a exploding star. Eventually we will discover every atomic particle and isolate every gene responsible for various human traits.

Those looking for similarities in chimps, for example, to humans, should equally be prepared to look at the *dissimilarities*, such as the pelt, the skull, jaw, teeth, hands (in the different thumb), the feet, the wrong hip joint preventing upright posture, the central nervous system, the chromosome pattern, the tiny brain and lack of intellect, the total absence of creativity and feelings of destiny or any drive to achieve *anything*. In short, a most unlikely candidate to have bequeathed to us the remarkable brain.

If mankind is a product of creation by alien beings, wherever it may have taken place, it could not be viewed as a one hundred per cent success. If the aim was to simply nurture the Earth and maintain its rhythms solely and specifically, such an enormous amount of intellect would not have been necessary. Therefore, this indicates a creation of beings intended for much greater achievements, and therefore much concern would be apparent among them about human behaviour patterns.

The ultimate goal is much more likely to have been to create beings who would *ultimately become creators themselves*, and the moment we set about the so far only discussed business of terraforming another world for mankind, then the process will have begun and their ultimate goal will have been achieved. However, as said, a one hundred per cent success rate could not be said to have been achieved, and if the descendants of the beings who may have been the creators *are* now present in our skies, they will be well aware of this by simply observing our savage side in the incessant wars, crime and negative destructive acts which we bring upon each other and the planetary environment. The worrisome questions must be, what do ‘they’ plan to do about it and when will they do it?

If this genetic laboratory did exist and the ancestors of mankind were transported around the globe, the tribal nature and hostility would have been noted long ago, but perhaps it *was* expected to some extent and the 'creators' assumed that, over perhaps a long period of time, all the negative genes would be consumed. Successive generations of the created beings around the world would probably not be aware of the profound nature of their beginnings, but their legends would begin to take the shape of 'sky gods' and the theory that they were 'created', and these would become more and more distorted down the centuries, but would retain the central idea of creation. They would reproduce their ideas or thoughts on these matters in various ways, and all the cave drawings, artefacts, steles, inscriptions and engravings of people in awe of orbs, winged discs and spheres in worshipful poses, would gradually appear followed by all the legends, (when Moses wrote the Genesis account he almost certainly drew on much older versions).

If this cataclysm and loss of an advanced race of our ancestral 'hybrids' did occur and the legends of the initiators and teachers is true, then those surviving beings would be aware of the need to teach and enlighten the masses as quickly as possible, as they would be the last of the 'enlightened ones'. The human hybrid beings of around 10,000 BC were ready for their first lesson. They had completed the grand cultivation programme and their basic tasks of tilling the Earth with the sweat of their brow, but now their brain craved more knowledge and greater challenges. They may have felt that they could accomplish more on their own. Their Lords had taught them well the things essential for their survival. They knew the seasons, the calendar, something of the wonders in the sky. They had mathematics and geometry and had been taught the importance of hygiene, proper drainage, town planning methods, building techniques and a stable form of government. They were warned of Earthly instabilities and that buildings must comply to their strict instructions. If they complied, their buildings would still be standing for their successors to admire long into the future, even in the volatile regions.

All the great cultures and their impressive cities and edifices begin to rise, then the jealousy, the lust for more power, the territorial claims, the sacking of the, enemies' cities and vanquishing of the opposing armies begins. They had started to manifest symptoms of their negative genes, which would be apparent in humanity for centuries to come in their mental processes, and certainly manifest themselves in humanity in our constant wars and aggression today (are the alleged abductions and analysis of genetic material *now* explainable?)

And so, the great rising and falling begins and continues quite unabated for thousands of years. The negative forces still lurking within the brain have not been entirely consumed and begin to noticeably assert themselves and, although not *dominating* in the brain, certainly equal the positive qualities. Now, even the survivors of the holocaust that destroyed their advanced island habitat are dead. Only the hybrids remain, that were initially transported out from the creation zone to commence the Earth cultivation programme long ago, *and all the current earthly races could be their descendants.*

But occasionally, and strangely as the need demands, there is always that one being among them that answers the call, the one that stands out above all others, the one with just those vital additional extraterrestrial neurons that has the brain power to solve the problem. This little bit extra manifests itself quite adequately in some of the hybrids and the ones with the charisma, oratory and leadership qualities come to the fore. They come in all forms, from cruel tyrants to wise and enlightened rulers. The first civilisations emerge. There is a craving for order and control of the masses. Rulers and kings emerge down the centuries, but those negative forces are still retarding and holding back the positive. The old aggressive and suspicious side of human nature is just as strong as the orderly and the positive and now the eternal struggle beings. The conquering of other lands and enslavement of its peoples, the continual wars and strife and lust for power and territory continues unabated down through the centuries of human history. The unused cellular structure within our brains may not be developing at the pace our possible creators hoped that it would.

If an unearthly intelligence did bring about the human being, they would have long departed to other creative projects in the cosmos. However, they could not supervise the entire universe and, though their great memory banks would contain all the records of the negative and destructive actions on Earth as well as noting the positive actions perpetrated by beings *they* were responsible for, there would of necessity be a phase two of the plan, even if it *was* thousands of years after the initial creation programme. Even now some hypothetical alien discourse could be envisaged as taking place above our heads at this very moment about their 'plan of action'.

Whereas our thinking usually extends to what the next generation may be doing, theirs may extend into millennia and, as said, the primary objective may be for humans to become the same 'creators' ourselves and this will surely come to pass. Such beings who we envisage could have brought about our creation would also have some form of hierarchy, and leaders of their society would differ somewhat in outlook and in policies, no matter on how grand a scale they were functioning.

Just as we humans would ruthlessly cull a seal colony for reasons of logic involving the fishing industry and hence the general economy, these alleged cosmic creators may deem it necessary, from time to time, to cull some of *their* creations. Have unidentified craft always been with us observing the actions of their creations on Earth that they had hoped would be gradually eliminated by positive genes in their bequest to us within our skulls? Did certain human activity hasten the implementation of *phase two* of the Earth plan started long before by the aliens' predecessors in the first phase? Did 'they' see a direct need to destroy certain human settlements and the occupants in order to prevent them producing more of their kind?

If phase two of their plan was the 'emissaries', this would include the patriarchs from Abraham, Moses, Aaron and so forth, down to Jesus himself. As said in *The Angels of Abraham*, Jesus and *his* message clearly represented a milder and quite contrasting policy compared to the violence, terminating and 'genetic cleansing' of the Old Testament days. On Planet Heaven, or wherever our hypothetical alien beings that may be responsible for human creation come from, a clear change of leadership and policy seemed to occur with the disappearance of the hard-liners and the advent of a seemingly more 'softy, softly' approach emerging towards their activities on Earth, with the introduction of 'Plan Jesus'.

With all the direct contact with 'angels' in the Old Testament, who clearly did not see any risk to themselves in their close approaches to humans, here is another marked contrast with the beings *now* keeping their distance. Very *few* close encounters are reported where (except in abductions) 'angels' seek out contact with humans, and there is much wickedness to be observed in certain countries going entirely unpunished by any external unearthly source of 'wrath' today.

Section III
Are Humans 'Children of the Stars'?

1. Experiments?

If we think that the current evolutionary process of 40–50 million years to produce the level of intellect in a chimpanzee is slow, what of the staggering period of nearly 200 million years of the dinosaur epoch? After all that immense period of time, they were still munching foliage. Their evolution could only be viewed as indescribably slow, and if that were not enough they were snuffed out like a candle by the cold, detached violence of the cosmos.

Looking at the achievements of mankind in such a short period of time, the likelihood of billions rather than millions of years for the evolution of the human brain does not seems so strange. The question is how it got into our skulls? *Could* it have been a 'bequest' from an extraterrestrial higher life form? Are we the result of purposeful genetic manipulation to set us on the road towards becoming creators ourselves? Are we, on the other hand, direct descendants of cosmic beings that came to Earth over 50,000 years ago and suffered massive degeneration, as any group suddenly deprived of their culture, comforts and advancements would be? Strange myths exist in Earth's history which seem to indicate great advancement that ultimately declined or disappeared altogether. Even today, everything humans have achieved could be wiped out at a stroke if, for example, the broken-up comet that battered Jupiter came to Earth *instead*. For all our intellect and technology, we have *no* protection from such an event. Tantalising evidence of previous advanced cultures exists that were possibly wiped out by such events just as the ancient Egyptian priest hinted to Solon.

For all that, our solar system could be described as a quiet neighbourhood compared to the mayhem and violence occurring in deep space: great expanding gas clouds, supernova, even colliding galaxies. There must be many civilisations out there being directly affected by terrifying events. Are some planning to evacuate? To leave their system? Have they already done so? Did *our* ancestors arrive on Earth from deep space?

Where *did* the massively over-endowed human brain come from? Are the genetic similarities between apes and humans inevitable? Is the human body shape an evolutionary inevitability, with manual dexterity nature's ideal creation and common throughout the universe? Would a suitable 'womb', such as an Earth-like world of which there must be a great multitude, produce the human being given enough time? Are the laws on such worlds *universal* to produce the right type and number of amino acids for the process to get started?

Given all these cosmic catastrophes and wipe-outs could the human brain, with this massive evolutionary period and successive decline and advancement behind it, each time have an extra veneer of racial memory added to the last? With the advent of our astronomical science, it is sheer luck that we did not begin our analysis of local stars and find that the nearest one was extremely unstable and likely to become a supernova in a comparatively short time and, instead of being four and a half light years away, was less than one light year away.

If *that* had been the case, instead of ploughing all the enormous expenditure into the Mercury, Gemini and Apollo programmes with the ultimate aim of landing on the moon, would we have been frantically directing our energies and expenditure toward building space arks for some of us to escape planet Earth? Would secret lists of chosen ones have been drawn up and the masses kept in ignorance of what was to happen, with the prime object being to ensure the survival of the human race elsewhere? It would have to be well away from our system. The occupants would have to die in space rather than on Earth, but their *offspring* may ultimately reach another world. The only things they would know about living on an Earth-like world would be from instructional computers. This possibility of *our* ancestors being cosmic refugees will be considered in this section.

People of advanced years will be alive today who may recall stories of horror related to them, or which they may have overheard for their grandparents, of leeches being applied to people and the 'bleeding' of patients in the infirmaries (which seemed to be the cure for all ailments in those days), and of other remedies and 'quack' cures of the Victorian era, particularly using the new science of electricity. Such people will be among the best to judge the pace of today's advancement. Medical science *has* advanced enormously in a very short time. Many people hardly bat an eyelid now when hearing of new things coming into the field. They almost seem to expect it. Mankind is well advanced along the road towards becoming 'creators', and in another century we may have eradicated all known diseases.

Although growing human tissue and primary organs such as the heart will be achieved in the near future, thirty years ago this was still within the realms of fantasy or science fiction, and the proposals made around that time of extraterrestrial visitations to Earth and direct involvement in producing the human being were consequently quite brave and far-sighted. No doubt at the time they made many people consider another possibility for human origins, just as Charles Darwin's theories did towards the end of the last century by upsetting the religious factions. Such theories become less and less fantastic with each amazing discovery and achievement, particularly in genetic science.

Ancient man, watching the birth of his own kind and the animals producing their cubs, then competing with them for food, capturing and indeed defending himself and his family from some of them, had no earthly reason to suppose he was anything more than a refined or slightly smarter animal himself. However, even then the self-awareness and sense of destiny was flowering in the ancestral brain. Even then, the lack of written legends did not prevent the verbal handing down of creation legends that 'gods from sky' made the first men. Such myths as the North American Indian legend of the red man being made 'just right' in the ovens of the 'gods', while the white man and the black man (who they obviously must have been aware of) were 'rejects' as being undercooked or overcooked, are just one example of the countless legends of 'sky people' or gods 'making' men. Even the Bible is no exception, with the much-quoted 'Let *us* make men in *our* image', and the Garden of Eden creation zone. Surely it is totally illogical to accept the possibility of other intelligences existing in the universe and to expect them to be *all* less advanced than we are on our fairly young star system, towards the outer arm of the galaxy.

As said, *we* will travel to other star systems, *we* will find life forms and *we* will somehow become involved in the research and examination of them. It is hard to deny the possibility that it could have happened here on Earth in the past, particularly as learned people have admitted Earth could have been 'visited' in the past by extraterrestrial intelligence.

With the demise of the dinosaurs and the refoliation and proliferation again of all Earth's fauna, with the advent of the pongid apes not too long afterwards, any visiting unearthly intelligence would have found a true 'planet of the apes'. If such beings came to Earth *after* the apes had swung in the trees on a planet with no intelligent beings for millions of years, totally undisturbed by any human activity, then a short assessment of their total lack of the creative urge but adequate body shape, may well have prompted an alien decision by the life form study group on board their craft to make the famous remark, 'Let us make 'men' in our image'.

If this fantastic supposition is in fact true, then the descendants of the 'makers' would indeed show a prolonged and interest in overseeing human development for millennia and ultimately, when they finally felt we may be able to 'take it', the final revelations would be made to us and would almost certainly necessitate their appearance to us. Now, the long and patient observations of Earth, seemingly apparent to us by all the unidentified aerial phenomena observed for centuries, indeed millennia, begin at last to make some sense. Is mankind on the threshold of such profound revelations? The idea of extraterrestrials coming to Earth and cataloguing and experimenting with life forms and finally producing humans, *does* seem utterly fantastic. But then the things that our physicists are discovering in their deeply positioned bubble chambers and their research cyclotrons also seem quite fantastic, and they amaze even themselves with what they are discovering. Anti-matter and black holes are also fantastic, as are the discoveries in most other branches of science, particularly those in genetics.

However, for all this discovery, we have still not answered the questions about our mysterious past, let alone thought about our wonders of the future. Because of the obvious lack of heavy plant or lifting devices available to the ancients, it is only natural to look for some alternative explanations for their fantastic constructions and how they were achieved.

Perhaps we do underestimate the abilities of the ancients, but after many years of questions, explanations are *still* not forthcoming. The pyramids *still* astound us. It has been computed that to position the 2.5 million stone blocks making up the Great Pyramid of Cheops, for example, should have taken centuries, yet the Pyramids were supposed to have been built in preparation for the death of the ruling Pharaoh in *his* lifetime. How on earth did the ancients lift and neatly put into position a block 65 feet long and weighing 2,000 tons, such as there is in the still mysterious Temple of Baalbek in the Middle East, north of Damascus?[1] With all our modern technology, some of the achievements of the ancients would still be extremely difficult, if not impossible, for us today. How would *we* lift

---

[1] W Raymond Drake, *Gods and Spacemen in the Ancient East,* Sphere, 1976.

a 2,000 ton block of stone? Even if we do try to resist the tendency to assign any extraterrestrial connotations to edifices and constructions the ancients have achieved, what do we make of the extensive ground markings purposely made to be seen *only* from the air? The well-known markings on the Nasca Plain in Peru would *still* be totally unknown if we did not possess flying craft of any description, and therefore it seems logical to assume that they *were* made to be seen from the air.

In another work, I suggested that if any indigenous natives of the area of the Nasca Plain *had* witnessed extraterrestrial beings analysing life forms such as monkeys, birds and spiders and had noticed certain markings on the ground that the 'visitors' may have made as reference points (perhaps to compile 'Earth charts'), the natives may have reproduced them on a huge scale to encourage the 'gods' who had so fascinated them to return.

Nowadays, the usual reason given for the unexplainable is to assume that they are astronomical symbols relative to Earth cycles, growing and planting seasons and so forth. The general consensus seems to be that the ancients were totally preoccupied with growing, cultivating and planting, and that nothing else concerned them. Well it may be true, but the Nasca plain is a most unagricultural, unplanted, unfarmed, dry and rainless plain.

What did the ancients see in the sky that they were obviously so anxious to communicate with? It has to be said that the theory of unearthly beings even arriving on our world, let alone having any involvement in the emergence of mankind, remains as theoretical as evolutionary concepts, and equally as fantastic as the Old Testament version for human creation. Nevertheless, continual references are never made to 'our cousins the aliens', or 'our ancestors the extraterrestrials', and only circumstantial evidence, such as a ribbed boot print in stone in the Gobi desert and many other oddities remain on which to base conjecture.

With the 75 billion sun-like stars said to exist in the galaxy, not to mention the billions of others, it is not unreasonable to assume the existence of Earth-like worlds. Of course, this does not mean *75 billion* Earth-like worlds. The chances drop considerably when we consider the necessary conditions that must prevail on such worlds, not least their position within the exosphere or habitability zone for any similarity to Earth to be apparent.

Many systems may exist around sun-like stars with no life forms at all on them, whether they are favourably positioned or not, but even if we halve or quarter the number, we would still have upwards of 20 billion worlds as likely candidates for life. *Some* of those life forms would surely be advanced and intelligent far ahead of ourselves. The question must be, why haven't we received any intelligent broadcasts from them? The answer may be that we have already done so, but we were not listening at the time. Such beings (before our SETI programme began) may have moved on to a new segment of their sky and considered our system too young to have reached the technical ability to be able to receive their messages. We have been searching with our SETI programmes for thirty years and have scanned many frequencies. It takes a lot of time. 'They' may also be transmitting in ways totally beyond our comprehension. Furthermore, *'they' may already be here.*

It is perfectly possible that there may be beings with limited intellect that just carry on cultivating and nurturing their world with no moon in their sky and no comprehension of the twinkling stars being other suns and possibly taking no more notice of them than a chimp, a bird or a bee does. We could reflect on the fact that for nearly 200 million years, only lumbering huge herbivores and some ferocious carnivores walked the Earth with no species of high intelligence upon it at all. One wonders how many (if any) extraterrestrial visitors the Earth may have had during that huge span of time, although at that time other-world intelligences may have been getting on with their *own* evolution. After all, if we only assign a couple of million years to our own evolution, how many civilisations have arisen in such a huge time span with beings evolving to the point of space travel and possibly visiting Earth? What would they have made of such creatures? Would they have been seen as an outrageous freak of nature? Did unearthly beings have any involvement in the demise of the dinosaur population, perhaps to pave the way for other more intelligent beings? the dinosaurs appear to have existed in such numbers all over the world that they may have eaten themselves out of existence. How could the growth of the foliage have kept up with them? Nevertheless, the indications are that they met their end in a much more violent manner.

When we consider the millions of life forms on Earth, many probably not catalogued, any Earth-like worlds that may exist could have as many species on them as Earth but *all* completely different. No functional shape or form could discounted, but whatever the shape, surely their bodily makeup will be of the same stellar material that permeates space and that we ourselves are made of, no mater how it is arranged? Is the list of elements finite? Any makeup is possible and creatures depicted in the movies Alien or Predator are all perfectly possible.

We have a tendency to link technical advancement with intellectual maturity and the moving away from negative, destructive and wasteful activity. But *we* sit down and design starships and along with them 'star wars' *weaponry*, and our tendency for war and aggression has not diminished since the first violent act of one primitive against another. Yet we are transmitting into space to contact other beings. If we invite them here and they *do* look us over, *they may not like what they see*. However, as said, they may have covered our sector already.

Furthermore, their transmissions may all be going the 'other way', shall we say upmarket, instead of downmarket. They may consider it better to talk to older systems than to comparative newcomers like ourselves.

There is also the possibility that I mentioned in *Cultural Shock* that they may be a bit less naïve than ourselves and have found ways to blanket *their* emissions, while listening acutely for others such as the 'Earthlings' who have been betraying their presence so nosily now, for some seventy years to all the sun-like stars in the region that may have intelligent life-forms on their planets.

Hardly anyone giving the matter any thought, except perhaps some who choose to believe in a unique, divine creation of mankind and everything in the universe with *us* in mind, accepts the possibility that we are alone in this vast cosmic void. But it *might just* be so. A daunting thought, but no matter. Mankind will rise to the challenge. As may be the *intention* of a divine creator, we will go forth and multiply and populate the universe. We are already seriously discussing it when we talk of the terraforming processes for other worlds. All we need is time, and we have plenty of it.

It is estimated that our sun still has some 4.5 billion years of life left in it before we need to get too concerned about its behaviour patterns. And so we *may* be the first intelligent civilisation but as said, with the sheer insistence of the life force and the life particles known to permeate space, this seems unlikely in the extreme. And we know that intelligent life *does* exist in space. *We* ourselves on Earth are the living proof of this simple fact.

It has to be said that only circumstantial evidence, such as the massively over-endowed human brain and historical references and writings throughout history, exists to support the theory of unearthly beings being responsible for human development. Not one piece of hardware exists (unless the hangar 18 syndrome is based on fact, and small alien bodies do float in preservation fluid therein) that could be said to be 'not of this Earth'. Ancient (or modern) extraterrestrial astronauts may be just as mythical and tenuous a story as the preceding two sections. However, we cannot blame Kenneth Arnold for starting it all with his famous UFO sighting in 1947 because UFO phenomena and close encounters are quite ancient.

It may possibly be that, if there is any substance at all in the countless reports of alien close encounters, unidentified craft in our skies and alleged human abductions going right back to biblical times, and what those characters such as Ezekiel, really saw may all be part of a long and patient study of Earth and its inhabitants by generations of perhaps more than one type of extraterrestrial being. If we accept two simple facts, that the biblical patriarchs *did* exist and that they *were* truthful men, something quite profound *was* occurring in those times.

However, it is a fact that just as the ardent supporters of the Darwinian concept of human development from apes purposely 'aged' bones by chemical means and buried them to be 'discovered' later in Piltdown in Sussex, England, many ardent supporters of the reality of unidentified aerial craft seem in some cases to be equally desperate to have *their* theory believed and accepted. They have resorted to trickery such as pasting drawings of UFOs on glass, hurling pie pans and car hubcaps into the air, and photographing them both 'in flight' and by suspending them by lengths of cotton, producing many convincing and previously accepted photographs as 'unexplained' phenomena.

It seemed to be sufficient in earlier days for these cases to be accepted as genuine if a photographic expert declared that the negative had not been tampered with. Today, with sophisticated photographic analysis and enhancement techniques, some of those once convincing photographs have now been declared as fakes. Nevertheless, enough convincing material and photographs remain to make it a subject worth investigating. With apologies to Winston Churchill, 'Never in the field of strange phenomena have so many reports been made by so many people from such a wide cross section of the public.'

However, although unidentified aerial phenomena, their origins and occupants, power sources, missions, and perhaps ultimate plans for the human race will be discussed, the main purpose of this section is to discuss whether extraterrestrial beings in the past had any 'influence' in the development of mankind in the first place and whether the astounding possibility could be true that humans are 'hybrid' creations.

Since the whole concept of the evolution and development by natural selection of humans still remains a theory, and the Old Testament version of human creation remains questionable in the extreme, it is perfectly possible to theorise that unearthly beings, highly advanced, particularly in genetic engineering processes, *may* have strongly influenced the appearance of the human form. The Old Testament under close analysis almost compels one to read 'alien' instead of 'angel' in which case unearthly intelligence *would* have created humanity.

Although we can freely discuss such a topic and look at the circumstantial evidence, if it was discovered or made known to humanity that we actually *had* cosmic ancestry or were somehow the result of an experiment by unearthly beings, then in spite of being able to discuss now the possibility lightly and theorise upon it, there would be a complete and utter shock. All our previous beliefs, teachings, history and natural history would have to be reviewed, reclassified or completely disposed of. This is what is known as 'cultural' shock.

However, like any other kind of shock it would in time be healed, but it is not easy to predict the ultimate reaction of Earth's races to such profound revelations. Although imaginative writers did exist in the recent past, such as H.G. Wells and, earlier, Jules Verne, who envisaged flights to the moon and even extraterrestrial beings, the contents of such writings were largely dismissed as amusing hokum. Just as the Bible speaks of 'angels' descending on a 'pillar of fire', *we have ourselves* descended on the pillar of fire on our own moon.

It is only with the advent of technology which resulted in our *actually* going to the moon, and also a rapid advancement in robotics, genetics, medicine and so forth, that we have slowly begun to speculate. If we propose such things as growing human organs in laboratories and shifting the atoms of matter around to recreate practically anything by changing its atomic structure (theorised in a recent television programme), then we must ask, has it all been done before, perhaps long ago? Are we simply relearning things that possible cosmic beings could have left us by bequeathing part of their intellect to us in the advanced portion of our brain? If the human brain is so developed and over-endowed now, what will it be capable of when *all* the cellular material is utilised?

As said, the Broca's convolution is the interesting area of the human brain, and is not possessed by the apes. In fact, the hominids did not possess it either, as is evident from examining the inside area of the skull finds. The so-called *Homo habilis* does not even present us with a skull, only fragments. *Neanderthalensis sapiens* has insufficient space between the base of the skull and spinal column for speech to have been possible, yet we have this giant leap to Cro-Magnon man from a creature seemingly doomed to retrogression. *Who provided the springboard?*

It seems certain that Neanderthal man was an evolutionary dead end, and this deepens the mystery of how Cro-Magnon man could have evolved from him. If Cro-Magnon man *was* the result of genetic experimentation by an 'other-worldly' intelligence, he would at least be considered a limited success. Limited in the sense that when closely observed for a long time, his negative actions of aggression and incursions into savagery from time to time would be noted and attributed to these traits remaining in the brain which are controlled by positive genes. However, his high intellect in comparison to that of earthly creatures would be obvious. Such hypothetical creators would not be infallible but Cro-Magnon man would have been viewed as a reasonable success.

The decision to allow his proliferation may have been made on the assumption that the negative neurons may eventually be all consumed in his future generations, with the extraterrestrial positive genes finally dominating. From modern day archaeological discoveries, Cro-Magnon man's intelligence was apparent many thousands of years ago and we find certain intricate artwork, before the later cave paintings such as those at Lascaux in France, of carved antler bones and 'straight' spears made mysteriously from the obviously curved tusks of Mastodons. Some of the aforementioned carvings seem so intricate that they could only have been produced with the aid of a jewellers' magnifying instrument. What process was used to straighten the tusks? Clearly, the one attribute that stands out so obviously in Cro-Magnon man was his highly developed *creativity*.

There is also evidence that they had a high level of ability in clothing manufacture, defying our assumption that all beings of such a time period sat grunting in caves – an impression much encouraged by such scenes in our ancient caverns and theme park displays devoted to human life in times past. Why would so many earthly legends have the same theme as the 'making' of men in the beginning by 'sky gods' from the stars if there was not at least a grain of truth in it all?

Even within recorded history, it was thought that the stars were holes with Heaven's lights shining through, so it seems that the ancients would have not made the connection that those far off, twinkling points of light were other suns, let alone imagining other worlds in orbit around them (which still remains unproven today). However evidence for large planets does seem to exist, and the Hubble Space Telescope will no doubt eventually settle the issue.

Just to consider the possibility that advanced extraterrestrial beings came to Earth thousands of years ago and involved themselves in experiments of a genetic nature with Earthly life forms seems to be utterly fantastic, but for all that, the possibility *does* exist. Clearly, they wouldn't get it right first time, whatever they were trying to achieve, and this could account for the 'disappearing entities' mentioned in Section I who died out for no good reason and did seem to possess more refined attributes than the true pongid apes. The amazingly rapid appearance of humanity contrasts greatly with the enormous time periods for the evolution of other creatures.

As said, during the long reign of the rather pointless dinosaurs, other-worldly beings could have been evolving naturally to a highly advanced level. However, natural evolution was very probably responsible for some kind of branching off from the pongid apes at some time in the past, and, given enough time, may have struggled slowly to produce a human-like being. But the strongest circumstantial evidence for external influence on the appearance of our current body shape is the massive leap forward over 30,000 years ago and the rapid development of intellect. Inherited racial memories could be a bequest of extraterrestrials, giving mankind a feeling of ultimate destiny in the stars, a destiny that we will undoubtedly fulfil.

Further to this, the evidence in Neanderthal finds seems to indicate a being most unlikely to have been a predecessor of Cro-Magnon man, especially as he was, in any case, seemingly doomed to extinction through *retrogression*.

If such a fantastic experiment did occur and humanity *was* the result, we certainly would not be considered flawless beings, and continued analysis and involvement in human life, including frequent abductions, *would* occur. In spite of our own self-esteem, 'they', the descendants of the possible creators, may view us as currently functioning well outside of the intended parameters with our continual wars, aggression, crime and unsatisfactory behaviour patterns. Clearly their biggest problem would not be *how* to cure our negative mental traits but how to secure *our* co-operation or how to forcefully bring it about.

Today, with our current level of technology and cosmic awareness and lack of religious belief in comparison to biblical times, 'they' can no longer freely move among us masquerading as 'angels'. However long they may have been observing humanity, there would always have been an area on Earth where humans were busily slaughtering other humans, and a simple neurological operation by them may cure this human aggressive tendency immediately. An analogy of their problem with us may be Dr Frankenstein chasing his monster and trying to get him to lie down on the operating table for further 'adjustment'. If such a profound occurrence did happen, 'they' could not abandon us now to our fate as *we* are actually creating starship designs on our drawing boards and theorising suitable interstellar power sources. If such a possibility exists for human creation, *there must be a final phase,* but how could 'they' solve the problem of getting mankind to co-operate with them?

As said in the section on divine creation, even God could not be satisfied with His creations, and biblical teachings talk of Armageddon and the Second Coming, and a kind of devil or Satan, who is responsible for all questionable, negative or downright savage human behaviour patterns, being eventually overthrown. Both possibilities, i.e. divine creation and extraterrestrial creation, imply *further action* to be taken with regard to malfunctioning beings, with 'culling' or genetic correction methods perhaps soon to be used on us one way or the other, particularly as their abductions may show no natural tendency of positive neurons eliminating the negative ones in any progressive manner. As previously suggested, alleged and frequent abductions seem to be occurring which may be 'ongoing analysis' of human brain development.

Their mammoth task may ultimately be to win the hearts and minds of Earth's races, in order to convince us that *further* manipulation of our mental processes is necessary to stimulate the unused portions of our brain into action, and to eliminate all the negative, wasteful and savage portions of the neurological cell structure that serve no purpose and forever retard our qualifying to enter *their* domain. 'They' may see it as essential to carry out slight neurological 'adjustment' *on all human first born*, and the only way they could achieve this aim, would be through the complete domination and control of the Earth.

In spite of our obvious imperfections and shortcomings, the adjective 'awesome' is most suitable when comparing the human brain to other creatures, in particular our so-called 'cousins' the chimps. It is entirely possible that, given another 40–50 million years, the chimps may well have deserved the title of our cousins but the extremely short period of time assigned to the event, especially in producing an organ like the human brain, seems to suggest that *some* kind of advanced genetic engineering process removed many millions of years of slow, plodding evolutionary development in a comparatively short time. With what we know about 'time dilation', much may have been accomplished between missions of our 'creators' to Earth. If we dwell for a moment on the things we take for granted, such as the computer or the inside of a 'simple' electronic game or video recorder that *some* human being sat down and designed, it is not surprising that an organ could just possibly have within its cellular structure a bequest from unearthly beings who had quite possibly billions of years of mental evolution behind them, and this could be manifesting itself (partly) in the human brain.[2]

Earthly history becomes a little uncertain when going back more than around 12,000 years (to us at any rate), but any extraterrestrial intelligence that may have been responsible for human appearance would have an enormous amount of data in their planetary memory banks, including actions and operations on other worlds by their predecessors perhaps going back for many thousands of years.

Clearly, they were seriously advanced during their hypothetical actions on Earth, and if one assumes these events were carried out by their ancestral cosmic voyagers around the mysterious appearance of Cro-Magnon man, they have had a further 30,000 years of development since that time *at the very least*. If such beings could blend in with humanity by conversing, eating and drinking in biblical times, they would surely go unnoticed today.

It is not surprising that all the early projects to try and make some sense of the aerial phenomena adopted a 'cannot be, therefore they do not exist' attitude, as indicated by the late Dr J. Allen Hynek about projects 'Grudge', 'Sign' and 'Blue Book'. Professor Hynek was at first an ardent sceptic until actually getting to grips with the phenomena.

---

[2] Max Flindt and Otto Binder, *Mankind Child of the Stars,* Coronet, 1976.

I have purposely avoided trotting out a series of baffling UFO occurrences, but I will make an exception for a case that occurred as long ago as 1956, and is taken from *The Hynek UFO Report.*[3] It involved an occurrence in England where USAF radar operators at Lakenheath scrambled two RAF night fighters. One aircraft tired to intercept a UFO, that had been previously tracked at 4,000 mph by radar, hovering near Lakenheath then moving off at 400–600 mph. it made high-speed right angles turns defying inertial forces and wound up on the tail of the intercepting aircraft, until both out of fear and being low on fuel, the pilot broke off and returned to base, after climbing, diving, turning sharply and doing his utmost to shake off the 'bogie'. There have been *many* such reports.

However, with all these comings and goings, opponents of the theory are bound to ask for at least something substantial as evidence – some part of a ship, some artefact 'not of this Earth', or at least some evidence of high technology beyond our time.

Certain items have been put forward and suggested as extraterrestrial evidence, such as ancient batteries, ground lenses and sophisticated electroplating methods, and what appear to be gold aerodynamically functional and symmetrical artefacts resembling flying craft, but classified as 'birds'. However, such things could be put down to past risings and fallings of technology and our underestimating the achievements of our ancestors. Certain ancient Indian writings describing nuclear missiles *may* be pure fable and invention on behalf of the imaginative writers of such accounts.

---

[3] Dr J Allen Hynek, *The Hynek UFO Report,* Sphere, 1978.

Supporters of the 'ancient astronaut' theories could simply state that such artefacts which would be unexplainable *are* buried waiting to be discovered in the Earth, on the basis that if the evolutionists can get away with it, then so can they. They may also say that on the moon itself, unaffected by wind erosion and soil movements, evidence of past (or maybe still present) extraterrestrial activity are there to see. (I have covered much of this in my book *When the Moon Came* which is now being produced.) A certain NASA employee has stated, 'There are alien artefacts on the moon waiting to be discovered'.

To be sure, certain things that could be described as evidence of ancient astronauts are difficult for the 'debunkers' to explain away: the aforementioned ribbed-boot print found in stone in the Gobi desert by a joint Russian–Chinese expedition, or the fossilised human brain found in Russia from the carboniferous period.

Or the well-known machined cube with a groove round it and two convex ends. Actually the original, said to be made of high quality machined steel and found in a coal seam, is missing and the item now residing in an Austrian museum is a copy of the original. As I said in another work, perhaps the aliens came back for it. On the other hand, has some clandestine security organisation removed it and filed it away with other sensitive classified artefacts?

Although it could be theorised that cool advanced alien beings began experiments with an Earth species to produce *us* some 30,000 years ago, other unearthly intellects could have passed this way earlier than that and, perhaps, were responsible for initiating *all* life that ever existed on Earth. It is a process that we ourselves may be carrying out in the future, which would entail the introduction of masses of blue algae to terraform suitable worlds for our life forms. This substance exists in Earth's most ancient rocks over 3 billion years old and begs the question, could it have been introduced here on Earth when it was cooling from a ball of gas? As said, although blue-green algae is a primitive substance, it is *in itself* the product of an evolutionary process. Would conditions existing on the still-hot Earth have allowed such evolution, or was it purposely introduced in massive amounts?

However, the fossil evidence in the human evolutionary question seems to give credence to some sort of 'experimentation' going on, and perhaps they did use the creature 'Neanderthal man'. There is no good reason for his *natural* demise. *Was* it purposely engendered to make way for the successful Cro-Magnon man? Once we have isolated the genes governing height, build and intellect we may *ourselves* be able to create beings *to order,* as it were.

Were the 'giants on the Earth' in those days referred to in the Bible, a purposeful creation to fulfil certain tasks, then purposely annihilated after they had fulfilled their tasks? Even with our current rate of advancement, geneticists will admit to a future scenario when we may be able to construct any life form we wish, and this would include life forms 'constructed' for a *specific* purpose with 'giants' falling into that category.

Long into the future, when the exobiologists and geneticists that will help to make up the crew of Earthly starships encounter a planet growing wild and perhaps populated by beings of limited intelligence, they may just consider such actions as some envisage could have happened here on Earth. They would be a long way from home, and the time taken to shuttle backwards and forwards to Earth, even with power sources capable of super light speed, would be too much. Far better to analyse the life forms, study their genetic makeup and, perhaps, enhance their intellectual capacity *from organised bases set up there*. Whatever or whoever put them there, racial memories seem to be firmly planted in the human brain regarding the 'making' or creation of our species, and in order to accomplish this, some sort of bases or operational zones would have been required on Earth.

The Greek legends speak of an underworld creation zone, sometimes resulting in abominable beings that occasionally escaped from the nether regions to wreak havoc on Earth. Evidence does exist of vast underground tunnel systems all over the globe, some with accurately cut sides. What technology was responsible for them?

If they were the result of an alien technology that produced mankind, such beings would have noted our asteroid belt and the high probability of an impact on Earth from time to time, and this may well have encouraged the manufacture of such elaborate systems. If humans were 'created' by unearthly beings, although being quite proud of their creations, it is doubtful that they would intermingle with them and would certainly not consider them as equals. They would probably observe a kind of strict apartheid regarding them, unless becoming desperate like the gods in the biblical writings, 'who took unto themselves the daughters of men'. These may have been female abductions for experimental processes, as the Bible tells us that the 'giants in the earth in those days' were *direct results* of those actions in a union of the 'gods' and human females.

Just as there were those who grasped readily at the Darwinian concept of human origins as a viable alternative to the hard-to-accept pronouncements in Genesis, many people may also find it acceptable (or even flattering) to consider the possibility of unearthly intelligence being responsible for the advent of mankind, rather than a direct descendancy from apes. However, as said previously, our simian ascendancy would be acceptable to many if all the conjecture and supposition was removed and the bone fossils appeared on the table. (In any case, they would have no choice.) The uncomfortable part about it all at present is the 'propaganda' that seems necessary to support it. Having rejected *divine* sources for human creation the evidence *must* be found for the evolutionists have nowhere else to turn.

Many learned people, professors, exobiologists and so forth have proposed that the Earth could have been visited in the past by unearthly beings, perhaps many times. Having once said that and accepted such a possibility, one cannot then refute the situation and say, as some have, 'But that was in the past. The current aerial phenomena are probably all natural events which we know exist, such as ball lightning, which our physicists cannot yet fully explain.'

It just does not seem feasible that with the probable rarity of Earth-like worlds, in spite our impressive figures, extraterrestrial beings would simply arrive here, make a few notes and go on their way. They would be amazed at the enormous amount and variation of earthly creatures even in hostile zones, and most certainly when encountering such a fertile world teeming with life forms, their first actions would be to establish bases as remote as possible from the occupants and begin their lengthy and ongoing studies. If they arrived on Earth during the epoch of Neanderthal man, *he* may have been viewed as a candidate for the production of a taller, more intelligent and more refined being in order to bring his evolution forward in a giant leap across the millions of years that it would normally be expected to take. Once this process was firmly underway, they could have departed once again into the cosmos, onward to their next challenge, and the great memory banks on their home planet would have a new entry: 'Mission Earth: Life enhancement. Phase One'.

Naturally, they would return to duly check and monitor the progress and proliferation of their creations. Are we created in their image? Ivan T. Sanderson, in his book *Uninvited Visitors* mentions two gentlemen called Sprague de Camp and Willy Ley who did some real thinking about the ideal body shape for creative and constructive activities: creatures rearing up on two legs (being the minimum required for propulsive locomotion), freeing forward limbs for manipulation; eyes moving to the front of the skull to observe the actions of those limbs; the action of bringing food to the mouth to eliminate the vulnerability of putting the head *down* to eat; the proper development of manipulative hands and so forth.[4]

It has been suggested that, had the dinosaurs survived, a creature such as Tyrannosaurus Rex which was already bipedal, may have developed into a humanoid but reptilian, being. Our theorised 'creators' may have evolved over billions of years in a natural way to the seemingly ideal humanoid shape, also giving their brain a long and natural evolution, and by possible genetic enhancement of a suitable creature they propelled us rapidly into the future on to the road towards our own goal of becoming creators ourselves.

Why are there so many legends of men being 'made'? Are they mental manifestations of racial memory? If we assume the fantastic notion could be true, what kind of

---

[4] Ivan T. Sanderson, *Uninvited Visitors,* Tandem, 1974.

location would such beings have chosen for such an enterprise? It would have to be restrictive and provide safety from predators – to name just two considerations. *Would* a large island be suitable? Could the suppositions regarding Atlantis be true after all?

If alien beings did create us, wherever they carried it out, their creations' behaviour patterns would eventually be of some concern. *Was* the mission of the patriarchs and Jesus himself to modify such behaviour? If Jesus was an emissary as suggested in Section I, then considering the barbaric way in which he was killed, were the hypothetical alien creators tempted to give up on humanity as a hopeless case of failure for which they would be largely responsible? Or was his death foreseen and planned for? Were they able to administer special drugs to enable him withstand his treatment? One could envisage lengthy discussion on the alien world of the pros and cons of the operation.

On the negative side, humans had not been impressed by his actions and teachings enough to prevent them from killing him. On the positive side, it became evident after his death that many were quite convinced of his doctrines and ventured into other lands, at no small risk to themselves, to preach the message.

Furthermore, and more importantly, humans had been seen to actually forfeit their lives by horrific death as the prey of wild beasts, rather than renounce their belief in his teachings. Did this last factor finally convince the late emissary's controllers that the operation was at least a limited success? Did this, together with the obvious spreading of the doctrines and believers in their ever-increasing numbers, signify a fairly satisfactory conclusion to Phase Two of the overall creation plan?

As said, these beings that may have been responsible for human creation may think ahead in millennia rather than the next human generation, and if they did live for three or four hundred years, their own generations would be quite lengthy. The beings of each consecutive generation would only concern themselves with their own specific contribution to the *overall* plan of perhaps many thousands of years.

Was Phase Three genetic inquiry and analysis of the many humans abducted? Is it possible that today we may be in the closing stages of Phase Four of this overall plan? If their race is so advanced, and they are cosmic creators on a grand scale, our world may not be the only one which their highly advanced technology and creative methods have directly influenced. There may be other grand plans in operation for other worlds that help to fill the huge memory banks on their world.

We could ask, if they were so advanced as to genetically produce mankind possibly some 30–50,000 years ago, would they still be using similar methods today after all their presumed further advancement? The answer may be in another question: *is* there a finite amount of knowledge to acquire, and have they reached it? Most of the matter in the universe is hydrogen, and there must be a limit to the list of elements. Stellar material and molecular groups seem to permeate space and would be finite. Atoms and the smallest particles physicists can discover would eventually all be known. *Could* we eventually comprehend eternity, something going on forever, without feeling we are about to blow a neurological gasket?

There may be a point where beings would eventually become 'the enlightened possessors of all knowledge'. There would be nothing else for them to do but create, and the purpose would be for their creations *also* to become creators, such as ourselves. It may be that the most singular and frustrating difficulty, for all their assumed knowledge and advancement, would be the creation of a perfect and flawless being. Ultimate control of genes and their effect on behavioural traits may be their ultimate challenge and goal and to them be as difficult to realise as our search for the Holy Grail. It would be necessary to analyse human mental advancement over many generations. *Are* the alleged human abductions part of this quest? In almost every case, some kind of invasive surgery is alleged to have taken place and involved removal of male and female genetic material. *Is* it the prelude to the fourth and final phase of the plan? In one rather interesting and convincing case, a certain female related that she actually observed rows of foetuses in glass retorts, floating but alive in some clear blue liquid. Those observed foetuses may now be fully grown *hybrid* beings *ready for use*. The whole purpose of the abductions may have been to produce these quasi-human hybrid beings to be reared specifically for insertion into influential positions on Earth, rather like 'sleeper agents' awaiting the final call. They may have superior abilities to enable them to reach the height of their professions and directly steer world affairs towards their final phase, which may be their 'second coming' and profound revelations to mankind of our true destiny as creators and of the story of our own creation. With our heavy reliance on electronics in all our transport, defences and everyday life, they could bring the Earth to a complete stand still.

'They' may then explain the need, as they see it, to eliminate all the negative factors responsible for our darker deeds and actions which are constantly retarding our advancement by 'neurological adjustment'. (Hopefully on a voluntary basis.) *Is* it possible, as previously suggested, that the only solution (as they see it) would be for simple neurological surgery *on all future human offspring*, which they would rather accomplish with our acceptance and co-operation than in any other way? Is their final coming at hand, possibly at the end of the current millennium, the culmination of Phase Four and the end of the programme?

2. Were Our Ancestors Cosmic Refugees?

With Earth's turbulent geological past shown by marine fossil evidence on high mountains, these mountains may once have been sea beds, and no doubt some sea beds were once high ground. Is it possible that some clay seam may, in the future, give up some quite momentous discovery, such as a petrified but nevertheless recognisable, space vehicle. With such a tendency to cover up things, the authorities may hide such a discovery for a hundred years.

Is the reason for the highly evolved human brain due to the fact that our ancestors evolved long ago *elsewhere* and came to Earth as the descendants, born in space, of beings that left their home planet long beforehand and came to our solar system, many having lived and died in space before reaching it? An (Asian) Indian legend from the north-west frontier states that humanity descends from space beings who migrated to Earth and landed in the Lohit Valley.[5]

What is the *real* source of the following legends from the South Seas, from Raroia in the Tuamotu Group in the Pacific?

> In the beginning there was only empty space, neither darkness nor light, neither land nor

---

[5] Robyn Collins, *Did Spacemen Colonise Earth?*, Mayflower, 1975.

> sea, neither sun nor sky. Everything was a big silent void. Untold ages went by, then the void began to move [orbital insertion?] and turned into 'Po' [Earth?]. Strange new forces were at work [gravity?]. The night was transformed, the new matter was like sand that grew *upwards* [the landing] then 'Papa', the Earth Mother revealed herself.

It sounds just like an account of a landing on Earth automatically controlled by an interstellar ship's computer, specifically programmed to search out and land on a suitable life-supporting planet.

With regard to our 'construction', the bipedal body form may be common and may abound throughout space in any intelligent beings on other worlds in the cosmos. Natural evolutionary forces may have been struggling to produce such a being up to the advent of the Neanderthal species, and this could have been the time when the great experiment began on Earth. However, with regard to human cosmic refugee hypothesis, the scenarios depicted in certain films, such as *Lord of the Flies*, may be quite accurate in portraying just how thin the veneer of civilisation really is in mankind. All vestiges of it could be stripped away in a very short time. There would be a reversion to savagery, and survival of the fittest would then prevail unless extreme discipline, firm leadership and a co-operative attempt was made to maintain some standards in behaviour and civilised existence. The will to survive is very strong: even cannibalism has been resorted to by marooned air crash survivors.

In the situation of an air crash or a shipwreck in remote regions or on a desert island, the survivors, though realising they were degenerating, would be aware of life going on normally in civilised areas and would retain the hope of rescue for quite some time and not accept their situation as permanent. Of course, this assumes they would be adults and not boys who ultimately viewed their existence as one long game in the aforementioned *Lord of the Flies* away from any constraints applied by adults.

If, then, we consider the possibility of ancestral cosmic evacuees who long ago left a threatened world as desperate refugees, surely the higher the civilisation they had left behind them, the worse they would be when suddenly deprived of it. If they departed on an interstellar craft, with computers administering to their needs, this would only be an extension of the facilities available on their own world. When arriving on the 'computer selected' planet Earth over 30,000 years ago, how would they fare? They would know so much yet be capable of so little, and their civilisation would have to start again from the basic mining of metals. However their immediate needs would be paramount and would reflect the careful planning and, more importantly, the choice of the craft's occupants in relation to initial survival needs. What would such beings be fleeing from? Hardly a supernova of their own star. If they had achieved their long evolution on their former world up to the point of the death of their star, they should have long left it and joined the abode of the creators. If their sun was a hot blue star, hurriedly expending its energy, they would probably not even exist at all, as life may have failed to gain a foothold there. If it was a nearby, unstable star, their escape may have been due to the many years planning and working towards that goal since its initial discovery.

Comets, asteroids and even companion stars would all be plotted and known of by beings that have a long evolution on one world, and no doubt many contingency plans had also 'evolved'. The possibility also exists that in humans, the long brain evolution could have been achieved by our possible cosmic ancestors going through a process of rise and fall, advance and decline in different planetary systems. Perhaps even unbeknown to themselves, their ancestors could have experienced great trauma and threat on a world in deep space and may have had to evacuate their world to survive. They may have seriously declined, losing their history and roots. Here again the possibility arises of ancient advanced technology being wiped out in an earthly catastrophe of the past.

Could there be a grain of truth in the ancient eastern writings of India that seem to be describing nuclear weaponry actually being used, and flying craft and their construction? They would have been viewed as myth a century ago, but could be seen in a different light with the advent of modern day technology. Astronauts and people living in the Stone Age exist on the same planet today. Could a section of humanity have achieved such advancement only to have it wiped out by a cataclysm?

Evidence exists on the Earth of celestial holocausts in the form of eroded craters and also geological upheavals, floodings and catastrophes, especially during the significant period of 12,000 years ago. Could such technology have existed and all trace of it been wiped out, perhaps waiting still to be rediscovered when some island rises from the sea and gives up its secrets? How many still unclassified artefacts and undeciphered writings exist in dusty museum basements?

The human brain seems to be old enough to allow for cosmic existence elsewhere (on perhaps more than one planet), with no knowledge of former ancestral adventures or trauma other than a firmly implanted racial memory of a destiny with, or previous existence among, the stars. Here conjecture reigns supreme. With regard to the floods of the past, were they caused by the impact of meteorites falling in the sea? Satellites reveal earthly impact craters and so some must have hit the sea.

It is also entirely possible that our ancient human ancestors came to this solar system long ago but that Earth was not the first world selected by their craft. During the 250 million years that it takes for our solar system to rotate once around the galaxy, the system may have passed through darker, dust-laden zones, when most of the Earth was uninhabitable or could have been frozen with a few life forms restricted to the more temperate regions. Such low temperatures may have prevailed for thousands of years. Given this huge time span for the human brain's evolution, the potential for past cosmic wandering, settling and surviving is enormous. Is the reason that we still have not rid our brains of the more negative and destructive genes the fact that we may have had to contend with this trauma and the struggle to survive, once again having to defend ourselves against all kinds of savage creatures?

Are the dreams everyone has of falling and waking with a start, or running from some giant or beast and not quite getting away, all racial memories floating to the surface as we sleep?

These imaginary cosmic refugees may themselves have caused great ecological damage on their own world, perhaps by a runaway greenhouse effect or attempts to control planetary temperatures and weather systems going disastrously wrong.

Such beings departing a doomed world surely would not leave *all in one craft*. The computer may search out stars in the local star group and attempt in that way to increase the chances of survival of the group as a whole. sun-like stars exist just over ten light years from Earth, near neighbours in cosmic terms. If mankind came to Earth in the aforesaid manner, did others go to Epsilon Eridani, or Tau Ceti? They may have found a suitable planet there. Perhaps this other group found a far more hospitable and suitable world with conditions closer to those on their own planet. What if these beings soared ahead of those on Earth, who were fighting desperately for survival? The refugees to that world may have kept, or even buried, a complete historical record of some kind telling of their arrival and of the system the *other* group went to. Could the beings there be our ancestors' *other* offspring that are far more advanced, and came to Earth after the discovery of some hidden data, to see if the other ancient refugees had *also* survived and proliferated here on Earth?

Long ago they may have scanned our part of the sky and received no intelligent radiations, but if they discovered such buried data, they would want to know if their ancestors' other group *had* survived and proliferated. They could have arrived here in our biblical times. If so, there would be a very good chance that a war was raging on the surface and some nation was overrunning another, and in that case, they would see a need for 'direction' and would establish bases for a long period of observation. They would also see the advantages of a lunar base, with the one face looking towards Earth continually. Perhaps they have secret location under the sea or at the north pole. Frequent UFO reports seem to enforce this probability.

If this race over generations had observed our incessant wars, they *would* rule out contact for quite some time. They would see positive advancement and great intellects, but as well as the Newtons and Einsteins and other great brains, mankind has also produced Hitler, Genghis Khan, Atilla the Hun and Caligula, highlighting once again the extremes of the positive and negative forces residing in the brain. Abduction of humans for mental analysis would also occur in these cases.

When hypothesising on cosmic refugees arriving on Earth, we could ask if they *originated* in deep space, but came to Earth from a colonised planet in this solar system? The other two worlds in our solar habitability zone seem *now* to be quite hostile to our life form, but we cannot be one hundred percent certain that they were *always* like that. Given the vast time suggested for our brain development, we could have had our origins far from the solar system, before it even formed, and could have had millions of years of evolution behind us. Mars or Venus may have been the more attractive and habitable planets *until mankind arrived*. Can we be sure that we ourselves, by ecological processes in the case of Venus, or a 'war in heaven' in the case of Mars, did not lay waste to them by our own actions, by the same processes that our possible inherited racial memories are guiding us towards today by despoiling the Earth? These possibilities will be considered in the following pages. We are hardly in a position to state with any certainty what was going on in the past when we consider the tiny fraction of *one* revolution of our galaxy during which we have been around here on Earth, when the galaxy has rotated 60 times in the last 15 billion years.

In this regard, one rather worrying factor is that it is evident in the fossil records that a mass wipe-out of almost all species occurred on Earth 250 million years ago, and since it takes exactly that long for the galaxy to revolve, *we are now back at the same point*, and approaching the end of the millennium which so many seers and prophets of doom forecast as a traumatic time for us.

And so, for all the excess of advanced grey matter within our skulls, we could still have managed to have carried out these damaging actions and have finally arrived on Earth and begun the great rising and falling, degeneration and advance with always this unknown, unseen force seemingly preventing our mental advancement beyond a certain point. (Consider the wisdom of the ancient Greeks compared to the ignorance of the Middle Ages.)

As said, there is the possibility that these restraining forces may not have affected or retarded the offspring of the group that headed for other stars, and that they may be among us now, their brains having fully evolved in a way that ensures the elimination of such negative and destructive neurons that seem to be retained in the human cellular material and are continually ensuring our regression. It really does seem amazing that 2,000 years after the Greek thinkers many still assumed the Earth to be flat.

Put the case that we *have* cosmic ancestry ourselves, and given the possible lower temperatures that may have prevailed, the planet Venus was the best choice for settlement of our kind. We have already conjectured upon hypothetical scenarios regarding the reasons for having to leave our initial planet of origin, so for now we could assume *Venus* to have been chosen for initial occupation.

We smile now when we think certain elements of the lunatic fringe in the 'flying saucer' days of the fifties spoke about men from Venus, with their blue eyes and long blond hair. Then all we knew about Venus was that it had clouds and was similar in size to Earth. Today, with our space probes, landings, and Voyager and Pioneer trips around the planets giving us planetary information relentlessly, we know now just what a hostile place Venus really is. But was it *always* like that? With our noxious emissions into the atmosphere from the burning of fossil fuels etc. going on relentlessly since the Industrial Revolution, the deforestation and masses of concrete being laid daily all over the world for motor cars and new towns, together with our depletion of the ozone layer and other general ravaging of our planetary environment, we seem to be doing a pretty god job ourselves towards producing, eventually, a Venus-type environment right here on Earth. It is often said that history simply repeats itself. Everything repeats itself, even the simple styles of clothing we wear. At the other extreme, even the expanding universe may return again to commence another 'big bang'.

But to return to the subject, we are continually warned by science that we could eventually produce a runaway situation with regard to the so-called Greenhouse Effect, where heat will come in all right, but will be held in until the tide starts going out, *and* go up into the atmosphere and so on, right here on Earth.

So perhaps we *are* just simply ensuring that history, with regard to our solar system, *does* repeat itself. We may have done it all before in the distant past and are now setting things up to do it all again. What planet is next for the chop? Mars? After all, serious plans are already afoot to go there. It's only a matter of time before colonisation. So we ravage Mars. By that time, instead of theorising on the terraforming process, we may have actually achieved it on Venus. So off we go Venus, 'terraform' it for our life forms, and then start the process over again.

What a fantastic irony it would be if we do bring life to Venus and then find out from something we dig up that we did indeed start off there. Life could rattle backwards and forward, to and from the planets in the habitability zone until we leave our system altogether. Imagine we are back to the time of our *first* occupancy of Venus. We've had our last public warning by the scientists, everybody is noticing the temperature increase, coats are discarded in winter, masks are standard issue in every city, water restrictions have long been in force and, in secret, plans are afoot for evacuation. Chosen ones and their organisers are full of noble thoughts of owing it to the life force that had struggled for so many millions of years of patient selection to produce us. We must ensure our life forms survive. The technology is similar to that which we have today on Earth. We have sent our surveyors and probes. It looks like we'll just have to make the best of it. Earth's spin is so much faster. It's going to be colder, with a slight atmospheric difference – we'll just have to get on with it. This must be kept top secret. The people will be told the fleet of shuttles are to be used for an atmospheric cleansing operation to alleviate the rising temperatures and to dissipate the carbon dioxide.

Then the day comes and the chosen ones secretly depart. A moment's silence for those left behind, then they are quickly forgotten as thoughts of self-preservation loom large. Brief thoughts on the arrival chances of the 'ark' shuttles. How many creatures will survive? Just hope the ones we eat do. We do not know what we *can* eat there yet. If only the requests for more finance for planetary studies had been authorised. It will be trial and error. Some of the braver ones will die finding out for the rest. And what about the viruses and bacteria? We've paid so little attention to this world, taking it for granted that one day we would populate it, but using all our resources to try and save our own world. Now we are on our way and vital question have not been answered. It was sheer luck about that asteroid. At least we don't have those huge beastly creatures to contend with.

Then comes the arrival. One ship out of every two lost. They might as well have stayed at home. Numbers are now depleted. The rest manage to land safely.

Further casualties of epidemic proportions. Flu and common cold germs encountered for the first time. Many predatory creatures abound. Animals thought harmless, the same size as our 'rooga' that sat outside our dwellings on Venus, now attack us and tear us to pieces. We shelter on the ships. We landed between the sea and the dunes. There must be a melting period under way as each time the tide comes in further. We must abandon them. We lack the energy to continually move them. Something plunges into the sea towards morning, lighting up the sky. By dawn a black wall of water approaches. We batten down the hatches. Many carried out to sea that were away from the craft. We must leave the damaged craft and trek inland away from all this danger and mayhem. Nobody wants to start a family too concerned for their future. The elected ones keep saying we *must* – we are in danger of dying out altogether.

Getting to know what to eat now and which animals are hostile. All our creatures have wandered away or been eaten by Earthly predators. Our corral was ruined and the creatures we had penned up were all washed out to sea. We must chip into stone some record of our arrival and bury it.

Still nobody wants to start families. We may have to move and move quickly. They would hamper us. Perhaps later, when we're more settled. We don't know if we are in a geologically stable area here. It seems like a rift valley. We are going to trek south to find a home and settle down. There are creatures here a bit like us from a distance. They seem quite harmless, just foraging for food. Look as though they are starving. We will show them what to do. We have no technology for body reprocessing here. We have to dig holes and put our deceased in the ground. Well, here we are, right at the bottom of the ladder again. From here, the only way is up.

The planet Venus may not have been the initial home of mankind at all, but rather the planet *Mars*. Fantastic? Perhaps not. Science tells us that copious amounts of water once flowed on its surface. The fact could not be possible without a more dense atmosphere, and that implies living beings. Our current body shape may be out of all proportion to its original dimensions of many millions of years ago. Conditions on Earth may have produced our fairly tall muscular form simply with the extra gravity here on Earth, and our activity in defending ourselves. Where was the 'war in Heaven' the Bible speaks of? Was it between Mars and the mysterious Planet X that may have existed where the asteroids are now? It is where a planet *should* be. Those tumbling rocks were almost certainly part of a whole body at one time in the past. Now of course much of its material would be lost due to the impacts so evident on the other worlds and moons or gone into the sun.

The Viking landings on Mars were a brilliant achievement, but there was no doubting the disappointment when not one single creepy crawly moved across the viewer during the camera's scanning of the vicinity of the craft out to the Martian horizon. The scooping and analysing of Martian soil revealed it to be more like iron oxide than Earthly humus. Of course, the remote arm could only work in one vicinity, but it was our last real hope of finding any life forms, however small and primitive, unearthly but within our solar system. Now, of course, a Martian meteorite recently found on Earth is alleged to contain fossilised life. However the real facts will only become known when mankind actually goes there in fulfilment of the objectives now being planned.

We do not know how deep the surface material or soil that gave it the name of the 'Red Planet' actually is, and it may be that a manned mission and setting up a base there will be necessary before we can be in a position to get the real facts. When our astronauts get around to doing a bit of real digging, something may turn up from the lower subsoil.

Our sun is said to pulsate over long periods so, as well as envisaging cooler periods in dusty zones of the galactic swirl, we could also quite probably have periods of much higher temperatures. Long ago, perhaps our planetary system *was* much warmer and Earth *was* a hostile, oppressive, steamy environment. Mars may have been a far different place than it is today, with dust storm and craters everywhere, and huge volcanoes straddling the equator pumping out so much lava and 'iron' dust that they have changed the face of the planet in their area. Four of them straddle the equator and one of them, Olympus Mons, is the biggest in the solar system. The Viking landings seem to have raised more questions than provided answers, and it does seem hard to imagine a green planet growing in that strange hostile soil. But then, of course, we must remember that that remote arm has *literally* only scratched the surface, and the possibility exists that the Viking Lander only scratched up sterile volcanic dust from the Martian surface.

The most profound revelation by far is that it would appear that copious amounts of running water at one time raged across its surface. If this turns out to be true and the valleys and tributaries were not formed by other forces – for example lava flows – then the implications of this are enormous. Running water implies a thicker, heavier atmosphere. Without this pressure the water would simply boil away. Therefore, we have a situation where we could suppose that Mars was possibly more Earth-like than it is today. Of course, as well as asking what happened to the water, one would also have to ask what happened to the air.

We retain our atmosphere simply because the velocity of the air molecules bouncing about and colliding with everything (which we can hear if we put a cup to our ears and listen to them battering against our eardrum) does not exceed the escape velocity of the planet. So what happened to the air and water on Mars? Given that it did once have both, then it would be easy to imagine that life once existed there, perhaps as the ancestors of mankind.

Something very dramatic may once have occurred in the region of Mars and the hypothetical planet that is now the asteroid belt. Those boulders must have once been part of a whole body and been formed under immense planetary gravitational pressure. Small rocks cannot accrete from small wispy gas clouds. Of course a lot of the material of Planet X, perhaps the outer layers, would have crashed into Mars, the other planets *and* into the sun, and large chunks still come by from time to time to alarm us on Earth. It seems certain that the hypothetical planet was far enough away from Jupiter's possible gravitationally destructive effects.

Planets do not naturally explode. *Was* it the result of the aforementioned biblical 'war in Heaven'? If the hypothetical Martian beings won the war by destroying Planet X (perhaps accidentally), the resulting impacts of planetary rubble would have rendered their own world ecologically uninhabitable. Perhaps the impacts weakened the crust and caused those immense volcanoes to spring up and further pollute the already thick, dusty atmosphere, maybe even pushing the planet's orbital track into its elliptical path and further exacerbating seasonal change.

Perhaps even the very atmosphere was affected to such a degree that it began to dissipate. Plans would be immediately drawn up for escape of at least *some* of their chosen ones to ensure the continuance of their kind elsewhere.

So, our hypothetical Martian ancestors find themselves in the unenviable position of having to decide who will be the fortunate ones chosen to leave and ensure the continuance of the 'life force'. No doubt they also would have some of their number travelling to another nearby star group. Hospitable planets being rare, hypothetical cosmic refugees would land on any planet that they felt they could survive on and in this case, beings could experience problems with the star of the planet they chose reaching the end of its life during their probable degeneration and re-emergence to an advanced space travelling race. After the dinosaurs' demise, our sun was *already* halfway through its life when the pongid apes were emerging. Could a cosmic refugee group have arrived at any time in earth space? Did one arrive millions of years ago, perhaps from deep space, maybe fleeing from a supernova threat as a relatively near star became unstable or approached the end of its life? One could imagine a fleet of starships arriving, say, 65 million years ago and the commander and his followers agreeing their search is over. 'This world will do nicely, thank you.'

There is only one problem: these huge fierce creatures lumbering about everywhere. They could not survive alongside them. Then, the commander addressed the ship's occupants. 'We have come a long, long way. It is known that planets like this are rare indeed. We need it to survive. We cannot pass by in the hope of finding another and, reluctant as I am to suggest it, the fact is that *they* survive or we do. I vote that *we* do. All opposed?'

Naturally, very few alien hands would be raised.

So, the decision having been made, off they go to the asteroid belt to direct a few choice hunks of rock towards Earth and let planetary cooling do the rest. Of course, if that event was followed by occupation, we would be most certainly finding some remains of them in the way of bone fossils. But the worrying thing for us must be that this type of scenario could happen today, with *us* being viewed as the irritating occupiers they couldn't live alongside.

When the asteroid belt was mentioned as possibly once being a planet, can one imagine any way in which small chunks of rock would 'accrete' or come together by the small gravitational attraction of gas molecules. I am positive they were once part of a whole body. I will simply call it a gut feeling. Chunks of rubble that come to Earth *are* probably part of that wandering rock pile and, although some are dated as older than three billion years – in other words, almost as old as the Earth itself – it is a fact that we can find rocks almost as old as that on the Earth. It is logical to assume that the *lost* material of the hypothetical Planet X would be the outer, younger layers that have long ago crashed into the sun and the other planets and moons, and it is the *older* material that now exists in the asteroid belt that gets knocked into Earth's collision course from time to time, so meteoric analysis *would* only show very basic life traces.

For whatever reason or whichever way our assumed cosmic ancestors may have come to Earth, here they were on a planet ripe for the taking, the master race with dominion over all others. If our ancient ancestors did arrive from the cosmos, it is logical to assume that they would bring a whole host of plants and other life forms with them and, to be sure, legends exists of star people bringing certain 'food-growing' plants to Earth. Canadian Indian legends speak of the Corn Mother, a 'lady from the skies', who brought the gift of maize from the heavens. We hear the same story in the Egyptian God Isis, bringing red and white barley and wheat from the 'heavens', even stating the star system as Canis Major (8.7 light years away).

Perhaps then, some of the creatures of the Earth who seem to remain 'each unto their kind', also have their ancestry in the cosmos as well as the food-producing plants we see all around us. It is stated that anthropologists working in the Olduvai Gorge, Tanzania, found several hundred extinct animals new to palaeontology, along with the famous Australopithecine finds.

But however or from wherever our hypothetical cosmic ancestors may have come to Earth, they would obviously in such a long time period, have suffered massive degeneration and reversion almost to savagery in their desperate struggle to survive so many obstacles. As said, the choice of professions of the ship's occupants would be of extreme importance.

Clearly a carpenter would be a useful chap to have, as would be a toxicologist, an agricultural expert, a botanist, a mining engineer and people of similar professions. But the need for strong leadership, discipline and organisation would be of paramount importance.

If such a group of cosmic refugees arrived on Earth a couple of thousand years after the demise of the dinosaurs, when the Earth had 'regenerated' after all those hungry mouths had ceased to strip the trees of their bounty as soon as they had produced it, the Earth *would* be like a Garden of Eden to the fortunate band, with fewer predators to bother with.

They would become lethargic, indolent, downright lazy and carefree, picking areas clean and moving on as the mood took them, and very quickly lose all interest in all their previous professions. Is this the reason (in racial memory terms) for the nomadic tribesmen that exist today on Earth? What is the root cause of cannibalism? Did the deceased contribute to the living on a former world? Body forms processed into protein wafers or bones ground up for fertiliser?

On the other hand, if they arrived when the Earth was highly populated with sabre-toothed tigers, bears, wolves and other carnivores, and if they had volcanic activity and geological upheaval to contend with, then they would be more alert to life and its perils and would have to use their brains to flourish and survive.

Their extra intelligence would be used to full advantage. It would be their 'edge' that would compensate against other creatures' abilities such as their keen sense of smell, ability to see in the dark and their being much more fleet of foot. No doubt the planetary environment would affect greatly the appearance of the refugee band as time went by, and eventually, they would change beyond recognition from their former selves.

Could something profound indicating cosmic ancestry be discovered in our mountain areas? Given the amount of time since they were sea beds, perhaps some very interesting discoveries could be made. But how on earth would we know where to start? It would be quite a different proposition compared to (for example) the finding of the ancient city of Troy. At least there were some writings of it and hints of its location for the archaeologists to go on.

Of course, when certain odd artefacts are found that are hard to classify, items that rock the boat a bit, they are neatly found a home by being labelled as possibly something they are not, just to get them out of the hair of the person doing the classifying. An example that seems to be most often quoted is a golden artefact shaped like an aircraft and proved to have aerodynamic features which was supposedly found in a Colombian tomb, but because aircraft did not exist in the ancient past it is a bird – end of story. To be sure, it seems certain that aircraft did not exist then, but was the artefact already quite ancient when it found its way into Colombian hands? What are these strange 'vimanas' the old Indian texts refer to, and what of their modern sounding armament?

Of course, some artefacts found in the ground can quite often be produced by nature's forces acting in unison and are sometimes too quickly assigned to the ancient astronaut theory. A thin convoluted shellfish fossilised in rock can look like a modern day screw. These natural forces in equilibrium produced the Giant's Causeway in Northern Ireland, and natural wind erosion is capable of sculpturing rock to look like something 'constructed', as in some parts of the American desert regions.

However, with regard to a possible significant find, and given the alleged tendency towards 'cover up' said to be prevalent among government authorities, how would such authorities react if archaeologists realised that they *had* found something quite astounding, and the news broke or was leaked to some newspaper that it could not be 'of this Earth'?

A secret team of government scientists would almost certainly appear rapidly on the scene and assess the find as quickly as possible. With the stark realisation of what it really was, there would be an immediate news blackout and 'D' notices handed out to the dailies with the cover story (possibly) that it was a lost satellite that the authorities had been searching for, that there was a danger to life through radioactivity, and that everyone should immediately clear the area.

This would be quite effective as, to be sure, the very mention of radioactivity *would* clear the area. The object would be unearthed and whisked away as soon as possible to some top secret remote location, a sort of Hangar 18, where the best brains in the land would descend upon it for lengthy analysis.

The cover story would be maintained, perhaps indefinitely. At first such remarks as: 'We cannot release this information without adequate preparation of the public for cultural shock and possible social disorientation,' would be made. Then, later, when the full horror of what it would entail had been realised, especially with regard to our established history, background, teachings and religious factions, not to mention all the current evolutionary theories of human beginnings and world history, we may *never* get to know about it. As well as the possibility of cosmic beings having arrived on Earth in the past, such beings could at this very moment be heading towards the Earth in a fleet of starships looking for a home. Clearly they would need to travel far enough away from their own system to avoid (for example) a supernova reaching their new home, and on this basis they could be here... well... tomorrow.

Of course, if they *do* travel anywhere, it is likely to be towards newer, younger stars than inward towards older ones, and another possible supernova. Besides, they may meet their match that way and couldn't expect to be able to take a world by force, if they had to, against a civilisation as equally advanced and perhaps as powerful as themselves. So they head towards the younger stars, towards our neck of the woods: a good, steady old sun, with plenty of life left in it and beings that may not give them too much trouble.

When they encounter the beautiful globe that is Earth, it is not very likely that they will pass it by and go looking for something better.

Human beings know from their own experiences about vanquishing other nations and overrunning their cultures in the search for what they want: the conquistadors brushing aside entire peoples and their way of life in the gold lust that was upon them; the 'invasions' of Australia and the American West. We know just what happens to those less advanced and on the receiving end. To be sure, we Earthlings would stand little chance against a determined race, millions instead of hundreds of years ahead of us. Although by probability Earth-like planets must surely exist, they may only exist as a maximum of *one* per million systems if our system is anything to go by. Given all the current and past UFO sightings, we could be forgiven for asking, 'Have they already arrived?'

3. What are UFOs?

It is quite interesting to reflect or consider the many and varied explanations given for the ongoing aerial phenomena in our skies. For example the contradiction that states, 'They are natural phenomena that science does not yet understand, but further investigation of them would not advance the cause of science'. Or 'They are from another dimension or a parallel universe'; 'They come from under the sea'; 'They come from inner Earth'; 'They come from outer space'; 'They are future human beings who have conquered time travel and have come back to a certain time in their past (our present) to study their ancient pre-history'. This last supposition is not supported by the fact that if they are doing so, they are taking an awful long time about it, given the lengthy time period of UFO activity.

Then there is the case that they may be the descendants of beings that were the original 'creators' of mankind. Also, there is the suggestion that they may be a phenomenon of the mind that only allows certain people to see UFOs, ghosts and other apparitions. Or that they are the advance party of an extraterrestrial race surveying Earth for a possible take-over. (Again, given the extensive time period of UFO reports and aerial phenomena for centuries on Earth, they are taking rather a long time in completing their survey.)

It has been suggested that some of the nocturnal lights are pure energy or plasma produced by enormous friction in seismic activity, as they are sometimes seen just before or during earth tremors and landslides. It is said, 'They are a freak meteorological phenomena connected with the little understood oddity of ball lightning'. Whatever we may surmise, someone has suggested it. But one thing does emerge from it all and that is that the phenomenon is real and will not go away. It seems to demand an explanation and not to be ignored except by just a few seemingly self-funded enthusiasts.

With regard to their shape and general appearance, almost every type of UFO imaginable has been reported in our skies. What accounts for the massive variety of shapes and sizes? UFO investigators, when interviewing those who have related an experience, show the 'victims', if that is a suitable phrase, a chart of something like 60–70 different shapes and varieties. Just how many groups of aliens are observing us? And if it only proves to be one, why all the different types of hardware?

If they all represent different alien groups, how do they all get along with each other? Are we going to have another 'war in Heaven' as they fight over dominion of the Earth and control of us?

With regard to some of these alleged UFO sightings, the radar, visual and aircraft-located ones are indeed hard to dismiss, especially if they show their expertise at out-manoeuvring the aircraft sent to investigate and clearly show signs of being intelligently controlled. These reports only seem to show up in books on the UFO phenomenon and one rarely sees any reference on any news programmes. With all the obvious ability and technology the air forces possess in enabling us to watch the Gulf War from our armchairs, surely one little UFO chase with a gun camera, following a weaving and escaping UFO neatly giving the airforce pilot the slip would not go amiss.

But maybe that's the whole point, the whole reason why the wraps are so tightly put on these events. No government would want to show its defence craft as inadequate to defend the country from something secret, perhaps possessed by an enemy, and they certainly would not want to admit that they did not know what was coming into their country's airspace and outsmarting them.

It certainly does seem, then, that when all the wheat is separated from the chaff, there still remains a hardcore package of sightings and encounters to be explained. However, just as ancient astronauts can be linked to everything if one tries hard enough, so too much exposure to UFO literature can tune one's mind to see things that are somewhat enhanced by one's imagination and turned into something they are not.

Of course, when one gets older and less impressionable and feels one has read all there is to read on the topic, the opposite becomes true. One gets cynical about the problem and looks for other explanations.

I can recall an incident when I was on active service in a hot spot in the Middle East, where terrorist activities were directed at us and one had to maintain and frequently check out certain remote posts where every stir and crackle is received with suspicion. Suddenly, my attention was distracted by something moving above me and I had to look up. A silvery metallic, circular object about four times the size of the full moon silently crossed my line of vision. My thoughts were, 'This clinches it, surely. They are real! Now after all that literature I have read on the subject, I have finally seen one myself.'

Then I heard a hissing noise, and looking to my right observed a chap pulling in some tackle at what I later found out was a meteorological point. I had witnessed the launching of a balloon, so often used as the reason for UFO sightings and guffawed at by the ufologists.

I knew what it was and was quite satisfied, but what would the nomadic Arab have thought, riding his camel away from the base as it passed over his head? He would have no way of finding out what it was.

If you drive along a tree-lined highway with a bright planet such as Venus or Jupiter low on the horizon, and flick your eyes from it to your car door, it will seem as though it is tearing along, keeping pace with you. Our eyes and minds can certainly be fooled.

But no matter how we explain them, there will always remain the stubborn, unexplained few that will not go away. Perhaps they are the scout ships, the real thing, the first of the many that will follow behind them to take over our world as their own is under threat from their sun. Of course, as with the case of our Venusians and Martians, it may be a disaster of their own making.

We must not talk as if supernovas were a comparatively common event, although there must have been many of them in our galaxy alone. Yet still no one (as far as we know) has yet arrived. This raises an interesting question. Are we the first? We know for sure that there *is* life in space – us! And we also know, as sure as tomorrow, that we will travel interstellar distances. What if we are the first and it is all down to us to go forth and multiply, to spread our life forms throughout the galaxy? Quite a mind-boggling thought. However, no doubt humanity would rise to the challenge. The way life is supposed to have started on our world would seem an inevitable process for any planet with the essential prerequisites of an atmosphere, liquid water and fairly steady sun. These conditions must abound throughout the universe and so the spontaneous emergence of life must be widespread throughout space.

To be sure, we have not yet even detected *for certain* that planets exist around other stars, but the odds against their not being there must be enormous.

It seems that many stars are binary and triple systems, but is this detrimental to life processes starting? Would there just be barren, cratered, or large gaseous worlds wobbling in erratic orbits about them, or are we to expect at least one Earth-like planet among them? Even our nearest star neighbour seems to be a binary system (a triple system actually with a distant third orbiting body).

Any extraterrestrial astronomers viewing our star from space would not see anything spectacular, but what they would receive would be the expanding bubble of radiations from our radio and television transmissions from Earth, which, from their initial emission, could be about 70 light years away by now. So any habitable planets or star systems within that radius could now be busily deciphering our broadcasts.

We may ask why we are not receiving any intelligent broadcasts from all these assumed intelligent civilisations ourselves? If they had the technology and intellect to broadcast in the first place, they would obviously choose the most logical and simple frequency to transmit on, say that of hydrogen, and we have certainly scanned this and many others in the SETI programmes. So where is everybody?

Maybe they are not as naïve as us. Instead of giving away their own position by uncontrolled emissions, they go out and reconnoitre for themselves, furtively scouting out the likely planets in the different systems until they reach us here. We may still ask why they have not made contact instead of all that flitting about showing us how clever they are. Well, we would not climb into a cage with a tiger, so why should we ask them to?

With our incessant wars and crime and our shoot-first-and-ask-questions-later policy, is it really surprising? We must be seen as a hostile and aggressive people, if not by our actions towards them, most certainly by our actions towards the planet and each other. If all these clever manoeuvres, shape-changing and even disappearing altogether are achieved by controlled craft, they must be enormously more advanced than ours. Yet there are those on Earth who can offer plausible explanations of electromagnetic power sources being their mode of propulsion, and the ability to become 'massless' and free of gravitational and inertial forces being due to 'anti-mass' field generators and so forth.

Of course, they may not be manned craft at all but remotely controlled units from a huge mother ship in high orbit. It does not really seem likely that they would tear about in flying saucers if they were advanced enough to bridge the interstellar gulfs so easily. Perhaps the occupants are some form of robotic or android beings. If they have been watching us since biblical times, they would have been a lot safer from us in those days, and could even have landed in front of biblical prophets. Now it seems certain that they have been fired on by Earthly craft, so it may be just as long again before they dare contact us, and with three millennia of further advancement one would expect great advances in their power sources.

In another of the books I have studied on the topic I read that the order had gone out from the military to try to force one down in order to learn the secrets of the propulsion before their potential enemies do. 'We must not have a flying saucer gap.'

In another chapter, I read of a secret and purposely constructed airfield with a special facility actually *inviting* them to land and make contact. What would the alleged aliens make of that – making futile attempts to force them down one minute and then inviting them to land in the next? Perhaps if they did harbour any doubts about us, that would make their minds up for them and convince them that they had made the right decision in *not* contacting us. Perhaps they do intend to eventually make contact and now no longer care if we observe their craft or not. It may even be part of the plan to 'condition' us and, when we finally stop chasing and shooting at them, they might consider it.

It is quite possible that these objects we continually see are robotic and originate from a lunar base. The movements and appearance of the beings in certain alleged abduction cases certainly suggests this is so. Perhaps the lights we observe moving about on the moon and the alleged symmetrical constructions there are evidence of their occupation. Maybe even the 'face' on Mars was constructed simply to give us something to think about. Most certainly, the moon, presenting one face continually to us, would be very convenient to them as an observation platform.

If we accept the biblical sightings as extraterrestrial phenomena – that is the 'close encounters of the third kind' such as Ezekiel's vision – then since that time their technology has moved forward a further 2,500 years. Now, it was pretty good to start with if they had interstellar spaceships, so if we assume they were enjoying a technological explosion then, just as we in our own small way are today, then they must be quite seriously advanced by now.

If we dwell on our current technology and pick out the things that are advancing rapidly, such as genetics, electronics, microsurgery, medical breakthroughs, communications, robotics and so on, it is just possible we might close the gap a little bit between us and them.

Perhaps that is what they are waiting for. Whether they will give us another 2,000 years is debatable, but maybe in just another hundred years we will astound even ourselves with our technological achievements.

'They' could be moving among us now – some of us may even be their genetic creations with their extraterrestrial brain power. They could very soon attain senior positions in all walks of life, particularly if they have mastered telepathic and mind-influencing power.

Some of the victims of the alleged abductions (in fact most) are females, and reveal astounding things under hypnotic regression. However, we must keep in mind that this treatment only brings out what the subject believes to be true (as also does the polygraph test).

Some female victims have revealed, under such conditions, that they have been surrogate mothers to alien offspring born after the implant that was carried out at a previous abduction.

Many cases of 'phantom' pregnancies occur throughout the world generally, which later on disappear. Perhaps it is time a few post-natal analyses were carried out on some of the cases.

Some of the 'abductees' question the aliens in their semi-delirious state and are communicated with in what appears to be a telepathic process. They are told that 'they' always know where to find their chosen ones, or as we might put it 'victims'. Some of the abductees relate that they sensed the aliens were inserting devices into parts of their bodies, in the skull or nasal areas. They even describe them as small, ball-bearing-like devices – a sort of homing beacon to allow the aliens to find them again.

At first, quite naturally, the females suffer a sense of violation but they are quickly subdued by a kind of mind scan that induces in them a feeling of tranquillity and even a sort of affection for the perpetrator. A kind of pattern appears in these cases with the descriptions of the alien beings. The beings that do the physical part of taking their victims are always small but the main man, the one who induces the tranquil feeling, is always taller, more reassuring, in control and almost a romantic figure. Many will see this as a typical human imaginative psychological manifestation. The 'Dr Kildare' syndrome. Alternatively, he may be a human-like being or specific genetic creation (or 'angel' as he would have been in biblical times).

If these hypothetical alien beings are capable of such things, they are not doing it just for fun. There must be an ultimate plan. How seriously do the clandestine security organisations, who surely must be paying some attention to these happenings, really take them? Given that there are great distances between cases of abduction, all following a similar pattern, it does not seem possible that each victim read the details of a preceding case and simply said that theirs took place on similar lines. Many abductions may still go unreported by the victims for fear of ridicule.

Security organisations would not be competent if they did not pay at least scant attention to it all. However their conclusions, if any, would not be for our ears but only for eyes of a certain branch of the MOD or the Pentagon. Clearly abduction victims who hide and suppress their encounters until hypnotic regression reveals them are not seeking notoriety or publicity.

Of course, we can watch programmes about such matters on television, which tell us nothing, make simple conjecture, show a few well-known clips and project them as probable natural phenomena, and seem slightly embarrassed about handling such topics. The victims of alleged close encounters often state that if they had such an experience again, they would keep it to themselves. No self-respecting airline pilot anxious to keep his job reports them any more, at least not publicly. A few of the victims of alleged abductions regret ever having revealed their strange encounters and naturally resent being viewed as slightly odd.

One wonders what the marvels of computer technology make of it all. Surely by now all the known facts could produce some kind of correlation, a pattern if not a conclusion. For all the assumed high intellect of the aliens, they would be unlikely to be perfect or infallible. They would have weaknesses and may be just as capable of folly as we are. They may become victims of some sort of ecological disaster on their own world, perhaps as previously suggested due to a their attempts to control or improve their weather systems.

They would, perhaps, have plenty of time to leave their own world in an orderly fashion rather than a panic exodus. Is outer space littered with alien corpses of those who have lived and died on the ships during their long journey? This kind of scenario is based on the level of technology we have reached on Earth, for it is perfectly possible that we ourselves could, with some serious effort, produce such ships for such an exodus.

They could be constructed in space for such an eventuality. Their shape would matter very little, as long as they contained some kind of shuttle craft to get the occupants to a planet's surface. Was the unidentified orbiting cylindrical device 'with arms sticking out of it' observed and photographed by James A. McDivitt from Gemini 4, just such a craft that perhaps broke up on entering Earth's orbit? Perhaps a better all-round design would be something on the lines of a huge space shuttle with the necessary aerodynamic capability and re-entry heat protection.

We could imagine beings arriving in the past, or indeed at any time. It is easy to envisage their joy on beholding such an inviting blue world after all the time they may have spent in space, seeing only stars and artificial light, breathing recycled air, drinking recycled water, popping miniaturised vitamin pills, only knowing through their educational programmes what planetary life was really like.

Among their many concerns of gravity, atmospheric gasses, bacteria and so on, their major concern would quite clearly be us, our reaction to them, and how they would be received here, for it would be sink or swim for them. They would never pass us by in the hope of something better.

We should not assume that these hypothetical extraterrestrials would be all powerful beings, with superior power or sophisticated weaponry. They may be gentle vulnerable beings, with wars and conflict totally unknown to them.

They could be like St Brendan's monks, coming ashore in a foreign land for the first time and leaving themselves open to the hostilities, or mercy as the case may be, of the occupants. To be sure, they would be the pick of the crop, so to speak: carefully selected beings of all the sciences and persuasions considered necessary to commence a civilised existence elsewhere.

Their first difficulty would be to establish communication to get permission to land. Once landed, of course, these types of extraterrestrial visitor would be entirely at our mercy. They would be heavily confronted by suited beings, monitoring and probing them, looking for alien strains of bacteria. They would have been directed to land in a flat, remote desert area. They would spend long periods in quarantine, with many different beings peering at them, carrying out tests.

They would have needles stuck in them, blood samples taken, skin scraped and poked for genetic material, be housed in a sort of hastily erected Hangar 18 and, of course, be fully conscious of it all and not humanely tranquillised or mentally calmed in the fashion of our hypothetical extraterrestrial abductors, as we humans simply do not have that kind of ability.

Then there would have to be lengthy Earth education programmes given to them, terms of settlement and protection from curiosity-seekers, robbers and all the usual unsavoury characters that abound on Earth. They could never expect to be left in peace. If their number increased they would be seen as a threat. If they did not they could be threatened with extinction. They would become like some form of super-protected species, or be viewed as freaks.

Eventually, they may begin to long for the return to the peace and security of their starship existence, sailing peacefully through the cosmos in their Mayflower of the heavens. It would probably take many generations of their presence on Earth before they began to feel settled or considered themselves Earth creatures.

Apart from a purposeful invasion of Earth, any other 'sudden' arrival of alien beings on Earth would only occur if they were desperate evacuees and had nothing to lose. As said, life-giving planets would be thinly spread and they would not pass us by. Although they may well be at our mercy, we would not immediately be sure of this and, if there were a small fleet of interstellar craft it could appear quite alarming to us and there would be the distinct possibility that we would attack to defend, giving ourselves the benefit of the doubt.

There would be immediate confrontation between the military and the science fraternity who would be vehemently against any kind of force, and the curiosity and desire would emerge to get a look at close quarters of the craft and the occupants. The military would probably be looking for ways to attack them, as the safety and responsibility for the country's defence is, after all, in their hands. They may suggest that the risk from alien bacteriological strains unfamiliar to us would be too big a risk to allow them to land. But, in any case, the final decision would be made by the politicians and *not* the generals. To be sure, the Earth would appear as Heaven to very advanced aliens leaving a threatening sun and a series of resource-drained worlds behind them. We could assume that they could subdue the Earth by force, but this is only if we judge them by our standards. If they had mentally evolved to having left brutality and savagery behind them, they would know of many other ways and means of accomplishing the same end, and that seems to be precisely what is going on today, when we theorise on the end result of all the alleged abductions and encounters.

Of course to save a lot of time they could say, 'Look, we need a home. We are going to stop the world for twenty-four hours. If that doesn't convince you of our power, we have other more unsavoury ways of making you co-operate. We will take Australia, thank you. You can even keep your coastal occupation areas. Don't bother us and we won't bother you. Here are your instructions for displacement of the current occupants. Signal us when you have completed them.'

Of course, there would naturally be the Earthlings' resistance movement that would be immediately set up. We would continually probe their defences and look for weaknesses. 'The human spirit will not be conquered!'

On the other hand, if we did accept the inevitability of their occupation and they did us no harm, we may learn much from them. Our technology may advance a couple of hundred years practically overnight. They may give us clean nuclear fusion without the dangerous waste products, suggest the firing into the sun of all our current buried waste using the space shuttle. They may tell us cheap ways of harnessing solar, wind and wave power, the secrets of advanced electromagnetic power principles, and even anti-gravity power sources. One wonders if any contingency plans or files *have* been opened on the subject. It would be of interest to know if such files exist, perhaps in the 'most secret' category, entitled 'Extraterrestrial Contact – Actions and Options'.

There must be many strange anomalies and artefacts in existence that clearly do not fit into the normal classification of other 'finds'. We know about the ancient batteries and ground lenses and what look like modern aeroplanes etc., but I mean something really serious. I do not mean to suggest that we really have the Ark of the Covenant or the Holy Grail, but some item must surely exist which is something quite unsolved and firmly and totally resists any efforts to classify it.

Perhaps with the new *détente*, or warming of the cold war, more information could come out of the East regarding a certain area in China near the Russian border, where a neatly buried row of strange thin creatures with small frames and large heads were supposedly found along with a series of strange stone discs with holes in their centres, rather like a music record. Apparently they bore inscribed letters or hieroglyphics emanating from the centre in an outward spiral, had a high-pitched ring when tapped and high cobalt content in their composition. The beings, when alive, were allegedly shunned and ostracised by the nearby locals because they were so ugly in appearance.

This all sounds a bit like the Chinese version of the Roswell incident, where small alien bodies were allegedly removed from a crashed disc in the New Mexican desert in 1947, near a USAF base.[6] The alien corpses are supposedly still preserved in some secret location.

On the other hand, alien beings may not be interested in subduing the Earth at all and may know of other Earth-like worlds encountered in their cosmic travels. Perhaps the Earth is seen as nothing more than a 'pit stop' planet, or they may view the Earth as a convenient oasis, newly discovered on the outskirts of the galactic desert, and simply add it to their list of known worlds where multi-cellular life forms have evolved.

With all the barren and crater-strewn, or cold gaseous worlds that probably make up the main bulk of all the supposed planetary systems throughout the universe, and in spite of our hypothesising about all the Earth-like worlds that should exist, in the immensity of space they would be spread very thinly indeed and may be likened to finding a snowball in the desert.

In this regard, it would be unthinkable for unearthly beings, encountering Earth for the first time with its abundance of life from the abyss of the oceans to cold hostile regions and the hot arid zones, to simply remark, 'How interesting,' and go on their way.

---

[6] Charles Berlitz and William Moore, *The Roswell Incident,* Granada, 1980.

There would be enough study time stretched out ahead of them to last for many generations of their kind to be adequately occupied in 'Earth observational studies', to probe, examine, evaluate, classify and catalogue, and for this they would require adequately remote and well-equipped bases from which to sally forth on their observation excursions. These bases would, of necessity, have to be in areas where humans rarely tread and the environment under an ocean or inside the Earth, perhaps from a polar entrance, would be adequately out of the way of at least the majority of human beings. Perhaps it is a factor indicative of their advancement that despite all the encounters (of any kind), all the alleged abductions, reports and photographs, we are no nearer to finding out any real facts about the UFO phenomena, and so conjecture rules supreme.

However in this section of the work, in order to make some sense of all the circumstantial evidence of the apparent interest in Earth and its life forms, at least since biblical times, it seems necessary to construct these hypothetical situations. These might seem acceptable enough if attempts are made to remain within the realms of scientific possibility and not to venture too far into what might be possible in the future.

This is sometimes not very easy, particularly when considering all the advances made in the field of physics and what those gifted individuals are finding with all those experiments in their cyclotrons.

Many theories, equations and supposed particles talked of in the past by such people as Albert Einstein are now being tested. Light speed was initially thought to be the ultimate barrier, but now even this is being challenged.

Scientists claim to have detected gas clouds in certain nebulae travelling at twice the speed of light. As previously stated, bubble chambers exist deep in the Earth to detect the muons and neutrinos that pass straight through the Earth and occasionally collide with the odd atom in the bubble chamber to disintegrate into other short-lived particles that leave their tell-tale traces in the liquid and can be photographed.

Particles are accelerated at enormous velocities around the massive cyclotrons to collide with other material. It is natural, and rather obvious, that any highly advanced beings conquering interstellar distances would be highly advanced in the field of physics and other sciences.

Extraterrestrial beings centuries ahead of us would no doubt be able to exploit, manipulate or convert any cosmic particles in existence. They may have mastered a technique of converting their ships and the occupants to 'tachyons', theorised particles with light speed as their *lowest* limit, and may be able to convert and transmit craft and crew to enormous distances at many times the speed of light. They may have for a long time harnessed anti-matter in gamma ray drives or even be able to dematerialise by using a form of anti-mass field generation. Imaginative science fiction has become science fact many times in our recent history.

What of the actual appearance of the alleged beings controlling the unidentified craft? Are they almost certainly robotic? Would the possessors of such theoretical advanced technology be present in the craft themselves, or only in the nerve centres of their alleged bases, which may well be long established in remote earthly or lunar locations?

What of the actual alien beings themselves? How would they look in appearance? There seems to be a sort of general pattern emerging from the abduction cases that have merited some investigation, particularly where hypnotic regression techniques were applied to the victims.

For example: large eyes, pale mushroom-coloured skin and hardly any mouth – just a thin slit with no verbal communication emanating from it. (Perhaps just a faint buzzing noise which may be the last vestiges of long unused speech.)

An emphasis is placed on supposed telepathic techniques. One gets the impression that the body shape is slight in relation to a larger skull and that this vague humanoid shape may be just how we ourselves may appear far into the future. Or if we do share a common ancestry with these alleged beings, perhaps we looked like them in the past and our planetary environment has grossly changed us to the body shape we have today. So what would the extraterrestrials that we surmise may share a common ancestry with us, really make of *us?* As well as seeing us chopping each other to bits on the battlefield, they observe our eating habits, see us slaughtering the other creatures that share the planet with us – peaceful creatures that we are not at war with – and then roasting them over an open fire and devouring them.

'No contact recommended with these beings for the foreseeable future. We must construct bases and laboratories for further study on their moon, under the sea and in their Northern Polar regions. It is hard to believe that we share a common ancestry with these beings. We must recover specimens for the study of their neurological activity.'

Nevertheless, in Earthly terms, we do have a high intelligence level and obvious over-endowment. It is even more strange that we as humans do not seem to utilise all our brain material, and have even more grey matter waiting to be used. During natural selection processes there is no storing up of spares – 'money in the bank', so to speak – or unnecessary over-endowment, so where did the extra material come from, and what is its purpose? Will we ever utilise it to the full? What will we be capable of when we do? How do we explain this excess unless it is a bequest by the ancient extraterrestrials.

The Einsteins of the world may be examples of our mental incursions into that unknown mental territory that one day will be commonplace to all of us. Mankind – if not the body then at least the brain – seems to be an anomaly in the normal evolutionary process, a baffling misfit. In spite of our most negative behaviour, we must be considered as an absolutely unique creature compared to others on Earth.

As all the alleged abductions seem to have at least one thing in common in that they all involve the removal of genetic material, someone clearly has more than a passing interest in us, and all the indications are that this analysis is searching for signs of human development.

However, even if they had not been involved in our original creation, they would certainly want to probe into our minds, try to establish our makeup and get to the bottom of our mental processes that cause us to engage in the activity of continually slaughtering each other and seriously ravaging the planet to boot, and then to display the ability to closely analyse every planet in our system, design interstellar craft and theorise about 'growing' bodily organs in laboratories.

Of course the abduction cases, if proven to be dreams or all in the mind of the alleged victims, could still have actually happened but long in the past. As every single case seems to involve the removal of genetic material, perhaps they are a type of 'racial memory' of the actions carried out by the 'creators' of the human race. Perhaps hypnotic regression brings them out from the mind's hidden recesses, in other words from the neurons inherited from extraterrestrial ancestors and containing *their* inherited racial memories.

It is entirely possible that there are people among us who may have carried out certain secret genetic experiments on apes, as the films and fiction stories suggest, but generally our typically human code of ethics would certainly prevent us from openly carrying out such actions, even if the law did not. But that will not stop the genetic experimenting proceeding towards the point when we are able to pronounce that 'we *can* do it'.

As many openly express the belief that we came from monkeys, we wouldn't be human if we didn't have some curiosity as to what a halfway stage would look like. No doubt it would be a formidable task to bring about such a thing. We couldn't produce one naturally, so some 'monkeying' around would have to take place.

Many human cadavers are donated to science, but do the medical scientists and biologists present the relatives of the donor with an *exact* list of their activities and actions with regard to the donated body? Of course, there must be a strict code of practice, but can we absolutely guarantee that *every single country* in the world observes it?

What about an ardent believer in the concept of humans descending from apes? He or she may donate his or her body to medical science with the express wish that an attempt is made to try and genetically produce a missing link, or halfway stage to prove the theory, using his or her body material. Of course it still wouldn't be allowed openly, would it?

During the Second World War, a body was donated to the military to be floated ashore as Major William Martin, drowned after an air crash with papers suggesting certain things calculated to fool the enemy. Of course, the noble intention was to save lives. No matter what the wishes were of the donor interested in the field of genetics, they could not be said to be saving lives, merely satisfying curiosity. In any case, a full body would not be necessary for such uses unless it was brain-dead on a ventilator and was allowed for testing reactions to other processes.

The scientists we think could be actively engaged in such furtive and unethical work could themselves quite easily supply quite enough genetic material for their purposes in that field.

Our alien astronauts would not necessarily possess this noble code of ethics and might simply carry on their activities in as cool and detached a way as our research scientists did in their experiments on live animals when producing all those hair sprays and cosmetics. This *was* obviously cruel but our noble code of ethics *does* retard our medical and genetic experiments.

The more we look and marvel at the human brain in spite of its equal tendency towards negative behaviour patterns, the more convincing becomes the theory that part of it had its evolution elsewhere, perhaps as said in the early part of this section on Venus or Mars. That enigmatic face on Mars, previously mentioned and photographed by NASA, has been widely publicised and caused quite a stir when first shown, especially as it is reported to be 'bilaterally symmetrical'. Where the pictures are of *that* I do not know. All I have ever seen is half a face, the other side continually in shadow.

It has been stated that NASA, has more recent photographs *but has not released them.*[7] To be sure, it is

[7] Don Wilson, *Our Mysterious 'Spaceship' Moon,* Sphere, 1976.

decidedly human-looking with eyes, a nose and a mouth but of course analysts, mindful of earlier disappointments of finding no canals, lichen or any primitive vegetation, no creepy-crawlies crossing the patiently watching camera of the Viking Lander and in fact no trace of life at all, remain decidedly sceptical – particularly as the close-ups of that world reveal it as decidedly 'moon-like' in appearance and almost as hostile.

For all that, the Mars missions were stupendous achievements by humans and once more seem to highlight the total improbability of our brain evolving here on Earth from apes in the ridiculously short time span allotted to such development.

To refer again to alleged lunar constructions and the face on Mars, we could assume there is always the possibility that they were constructed by extraterrestrial intelligence with not only a pleasant diversion for them in mind, but a clear message to us: 'Look, you're not the only intelligence in the galaxy. What do you make of this?' Only close inspection of the 'face' by a Mars landing will reveal the facts. Even if it does prove to have been constructed, it would hardly be possible that it was the work of the original hypothesised inhabitants. If beings did once exist there and produced an effigy in their own likeness for the benefit of planetary observers on Earth in our future, it would have been a long, long time ago and, given the amount of planetary surface erosion that must exist there with all those dust storms and so forth, it would be in a far greater state of erosion than say the Sphinx in Egypt. (Of course, the Sphinx's appearance wasn't greatly helped by Napoleon's soldiers using it for target practice.) In fact, by now the Mars effigy would probably be eroded away completely.

We know that it is a human trait of the eye and brain combination to make sense of things and make logical and symmetrical images out of chaos. This is evident when we see wind-eroded effigies in desert regions on Earth. There is this tendency to see them as castles or man-made constructions. Perhaps the same situation exists on the moon with all that jumbled chaos of rock. Some of them are going to fall into pattern with symmetry somewhere, *and they most certainly do*.

But for all that, one would not expect an entire book to be published on them. Nevertheless, one author spent enough time poking around the photo tubs at NASA to gather enough information for one, and some convincing photographs and sketches do appear in it showing symmetrical constructions on the moon's surface.[8]

Astronauts have allegedly commented on them and used codes to describe them as though another veil of secrecy had descended on information everyone has a right to know. Added to this, a Russian astronaut was reputed to have said, 'The public would be amazed if they knew what we have seen in space'.

The book dealing with the alleged lunar constructions also shows what the author called 'X drones' working the rims of the craters and, sure enough, symmetrical X-shaped artefacts were shown perched on those crater rims.

The trans-lunar phenomena (TLP) sometimes referred to include lights in craters and moving across the lunar surface. But of course, meteorite activity doesn't stop for the night and impacts would surely show up as white hot and growing for a while, and any equally hot material ejected from the crater would appear to be crossing the

---

[8] George H Leonard, *Someone Else is in Our Moon,* Sphere, 1978.

moon as it rose and fell before cooling. It would appear like a travelling light, assuming it is possible to observe such a comparatively weak light source with our optical devices.

TLP are not new and are almost as old as the telescope. Unidentified aerial phenomena are not themselves new. Are we fast approaching a culmination or answer to it all? Surely 'they' do not intend to observe mankind and the Earth indefinitely? If they are approaching the final stage, of some grand creation plan, will the end of the current millennium be a serious happening for us, or will the cases of UFO encounters continue to roll in well into the next century? Is it unusual for alleged advanced beings to display a sense of mischievous humour by taunting us?

As mentioned in the UFO event of 1956, cases are on record of unidentified craft quite effectively evading fighter aircraft sent to intercept or investigate them, and no doubt much interesting film footage exists under wraps of such craft darting away, perhaps taunting the pursuer or even winding up on the tail of the chasing aircraft. No amount of freedom of information laws would secure *their* release. They would be held under special category security classification for anything up to one hundred years, and would be assigned the protective phrase, 'Held for reasons of national security'.

Of course, the same classification regarding security would be applied to any hardware artefacts or hard evidence of any extraterrestrial objects found on Earth, and this kind of action by keeping the general public in the dark on such matters may save the authorities much embarrassment in having to admit that they do not know how to deal with it all. But it will ultimately work against them in any panic, horror or hysteria by the masses when sudden and alarming revelations of a profound nature may come upon them. Far better, surely, for some form of preparation plan. A far more sensible approach would be to introduce a gradual indoctrination of the masses over a long period of the possibility of there actually existing in our skies representatives of another civilisation not of this Earth.

There is not a country or an area on Earth where some form of aerial activity seemingly unrelated to Earthly and explainable causes has not been noticed, witnessed or reported on, including entering and leaving the sea, heading towards the North Pole, or being in the vicinity of, and actually on, the moon. The astronauts of all the space programmes up to their cessation with Apollo 17 have reported bogies, UFOs, large objects and moving lights or trans-lunar phenomena of one sort or another in the lunar region.

As said, the moon would be a perfect observation base and also a possible source of much sought-after minerals.

To return to the subject of Earth, a well researched and very convincing book was written on the alleged recovery of a crashed UFO in the desert in New Mexico.[9] It was apparently brought down by an electrical storm, the disc and its dead occupants being whisked away to some research hanger on a well-guarded air base. I suppose it is possible to have the technology to travel interstellar distances yet still be brought down by a lightning strike, yet our own rather unsophisticated aircraft frequently suffer lightning strikes (I have witnessed myself a series of small holes that had burnt through the metal) that allow them to continue flying quite safely.

The book has a reliable witness, a solid citizen and ex-army officer, witnessing the wreckage before the military

---

[9] Charles Berlitz and William Moore, *The Roswell Incident,* Granada, 1980.

arrived to hush it all up and clear the area and declaring he saw the alien bodies and unearthly craft. We could ask, does it seem likely that the US would be spending such large sums of money on the SETI programmes and other research to do with extraterrestrial intelligence if they had the evidence sitting right there in their laps?

But then maybe that's just what we are supposed to think. Perhaps SETI is a ruse, and part of the price they are willing to pay to keep it all under wraps and be the first to learn the secrets of the alien craft's propulsion systems. What an edge to have on any potential enemy. Of course, as the incident happened in a very sensitive area of secret flight tests and development etc., it would not be surprising if such hardware were quickly recovered and taken away by the military.

It seems then that there are many regions from where our alleged alien visitors could have come – nearby star systems, inner Earth, outer space, out of the sea, from the future or our own fertile imaginations,

However, it still remains possible that when our excessive, unsustainable population ventures forth to a terraformed new world, we will simply just be extraterrestrials going home. And, since masses of this blue algae occur in Earth's primitive soil, the whole business of terraforming may just be a repeat of such a process that was carried out long ago prior to the life process beginning here on Earth. As said before, until we leave our system forever, we may just be shuttling back and forth between the habitable planets near our sun, which would account for some of the huge amount of time required for the development of the human brain.

It is interesting to speculate what state of advancement we would have reached today if we had commenced our evolution at the first appearance of the dinosaurs. If we had managed to avoid being wiped out by them, after nearly 200 million years, they would still be just munching vegetation, but what (assuming the same rapid development of the brain) would we have achieved in all that time? We would most certainly have colonised every planet in this system worth the effort and probably those in nearby star systems also.

When we speak of Earth's turbulent past and all the cosmic threats to our world, 200 million years is a long time to be left in peace before the serious happenings of 65 million years ago.

If we had been around all that time we would have most certainly foreseen the event and prevented it from happening. Perhaps then, if we had not terminally and absolutely depleted their numbers, the dinosaurs, and particularly those like Tyrannosaurus Rex which were already bipedal, may have evolved into the kind of distinctly reptilian humanoid depicted by the Canadian Museum of Natural Sciences and featured in the book *The Search for Extra Terrestrial Intelligence* by Edward Ashpole (Blandford Press).

The creature was alleged to be biologically credible and had an air of intelligence about it. In fact, it was this creature that gave me the idea for the 'Arturans' in a work I called *The Second Coming*. The creatures of 'Artura' (now the asteroid belt) were accidentally wiped out by our ancient astronaut ancestors when they destroyed the planet once orbiting between Mars and Jupiter.

When we imagine how extraterrestrials would appear to us, it is easy to assume that they would be humanoid. In fact, it is hard to get away from that kind of thinking, which is strange considering the diversity of life forms on Earth.

To be sure, we seem to be an uncomplicated, unclumsy design, upright and mobile, suitable for building things, with arms, hands and fingers well-designed for manual dexterity. It would seem we have the ideal form for creating and building things. It is difficult to imagine something like an octopus flopping about on an alien factory floor, working on a spaceship production line (although it could hold quite a few spanners at one time!)

If we had all that aforementioned prehistoric evolution behind us, it is hard to envisage anything being seen as a threat to us. It is strange that today we possess the ability, technology and 'know-how' to put some form of orbiting planetary protection up there, yet no positive moves are made towards using this capability.

However, to return to the question of our possible origins and looking at all the options of how our Earthly life form came about (or arrived), we begin to look more like a coolly carried out experiment and not the result of descent from some sort of desperate evacuees. One feels drawn towards the plausible theory that our progenitors travelled the galaxy looking for signs of any intelligence struggling to evolve and feeling it their duty to promote intelligence, they created us in their image by genetically enhancing less intelligent creatures. They made us into fully upright bipedal beings with manual dexterity and the intellect to create and reason. To propel ourselves forward into a rapidly expanding technology by moving quickly through the basic stages, the discovery and mining of metals and ultimately becoming creators ourselves. The ability to mine, smelt, forge and create metal artefacts was the *first* giant leap for mankind.

There must be a distinct link between this alleged early boost to our advancement on the road to higher intelligence and all the apparent interest the so-called UFOs cause by their appearances, close encounters and supposed abductions. It would appear that a great interest is shown in us for some good reason other than the simple curiosity of having found life forms. To be sure, any passing intelligence or 'star men' would stay a while after encountering Earth and its abundance of life forms to study. It doesn't necessarily follow that they have anything to do with our development but, given our earlier equation in which we established that our brain has needed so much more time for its evolution, it would appear that part of our brain material has been around somewhere else long before being donated to humans and that intellect was given to us by the aforementioned 'genetic creating'. The facts do not seem to allow purely Earthly evolution to be considered. If this kind of experiment *has* been carried out with us, then at what stage are we allowed the right to know? But do we really want to know? When would we ever learn of it? When would 'they' decide to break the news to us? What indicators would tell them we were ready for such a revelation? Would they *ever* consider it necessary to do so? How would we react to it? What kind of cultural shock and social disorientation would it cause?

In spite of all our preconditioning by many years of science fiction movies, close encounter reports, UFO sightings and alleged abductions with supposed telepathic dialogue with extraterrestrial beings and close-up observations of them, we are still as unprepared today as we would have been in none of this had happened. Nor does it seem possible to know how to become prepared.

Some writers have speculated in various works in the past that Christ himself was an extraterrestrial sent down on a mission to direct human behaviour. His dress and manner (apart from the miracles) would be no different from the people of his day than an impressive preacher in a suit in Hyde Park would be to us. Perhaps before his termination he was intended as an essential part of the preparation process prior to any 'revelations' to us. They must have been tempted to give up on us completely when we killed him, especially in such a cruel, long drawn-out way, not to mention the prior cruelties inflicted upon him before his rather barbaric execution.

Of course, many abide by Christ's teachings and there is no doubt that a serious effect was made on many people at the time, and that today there are obvious changes in peoples' behaviour. Many people observe the Christian ethical code.

Perhaps it was the aliens that removed his body from the tomb and took care of the guards with their weapons in 'stun' mode. The effects of his revival may have been dangerous to humans if they came too close to him afterwards, and so they were warned to keep their distance when witnessing his resurrection.

His reappearance was clearly designed to confirm his followers' faith and certainly must have reinforced their confidence to go forth to many lands and preach the gospels or the codes in many languages, regardless of the dangers to themselves.

Since Jesus previously made it clear in what he said to some Earthlings that he was 'not of this Earth' and then proved it by disappearing upwards (probably to the waiting craft), many *would* believe in the 'Jesus was an astronaut' theory.

Imaginative science fiction? Fantasy? Perhaps. But it certainly all fits: the bright moving light in the heavens; the mesmerising by a brilliant light of a human female and the implant operation by the 'angels'; the special powers and miracles; his admitted statement that he was 'not of this Earth'; the inability of humans to kill him when so many others died on the cross.; the return from the dead, again with bright stunning lights; the dangers of approaching him; and the rising upwards. It couldn't be clearer to those who believe the extraterrestrial theory. Even in his youth Jesus was a 'wonderchild' and the tutors and scribes could teach him nothing. In fact Jesus lectured and enlightened *them*.

It does not seem likely that a normal human being could have created such a long-lasting impression and be remembered for such a long time, and have people actually die for him, follow him and hang on his every word. Who else has there been on Earth like that?

Of course there is Adolf Hitler, the Anti-Christ, with his 'miracles' of transforming chaos into order in Germany before his lunatic military adventures. He will be remembered for a long time. He had a great following. He had people die for him, following him and hanging on his every word. He was another extraordinary being. There was certainly something special about both of them – each at opposite ends of the spectrum of humanity's Jekyll and Hyde, good and evil. This evidence of constant opposition of positive and negative forces in equal proportions is locked up in our brains and manifested in everything we are involved in.

So, are the beings we assume may be watching us considering another being to send among us, a 'second coming' so to speak? Or will the second coming be them themselves, coming to impart the message to us and to tell us of their plans for us, especially as we seem intent to go off adventuring into the cosmos. They may see the end of the current millennium as the time for direct and positive action to curb the mistakes made in the creation of our life form long ago.

It would appear that a rather unique character named Nostradamus, the great seer and prophet of the sixteenth century, had predicted a great king 'coming from the skies' in 1999 and seven months. One wonders who that might be, and why from the sky? The prophesy has no religious overtones or connotations so we are we to assume an extraterrestrial connection? There were also some references to the Mongols which may be Red China. With things the way they are there, with suppression of freedom, possible arsenals of fearful weapons, secretive behaviour, suspicion of the West and the tensions and 'sabre rattling' in Korea, it is all very worrying.

The end of the current millennium sounds like an interesting, if rather frightening, time in many ways. Here we are, busily dismantling our nuclear capabilities together with Russia, while countries in the Far East and the Third World countries are busily building theirs up and seemingly preparing for something.

With regard to the strange things seen in our skies, it is often conjectured that governments are withholding much evidence from the rest of us. They are probably quite flattered by this because in the first place they probably do not know much to withhold anyway, and secondly it means they do not have to admit that they do not know what they are. If they did it would put their defence forces in a bad light and make them seem inadequate for the defence of the appropriate country. After all, with regard to the hard information, they only know what they are told. It is the victim of the abduction, or the fighter pilot that has watched the weaving target suddenly move from being in front of him to being on his tail, who know the real answers. They have had the real experiences.

Of course, the way governments could be in the know is if the extraterrestrials finally decide to make contact. They are not then going to talk to the farm boy down the lane and say, 'Take me to your leader'. It is certain they will be well aware who the leaders and many other people who hold positions of influence are, and this high-level contact could take place unbeknown to the masses.

If the extraterrestrials are here patiently watching us, then clearly they would have some sort of base, unless they operated from a huge orbiting mother ship which because of our all observation and active space programmes is not a likely option.

Apparently, large targets have been detected moving quite fast under water, and objects have been seen entering the sea as well as leaving it. There was even such a sighting recorded in the ship's log of Christopher Columbus, and of course many less famous seafarers have reported objects emerging from or entering the sea.

To return again to the planetary catastrophe scenario. Although it would be kept secret, it would be interesting to know if any kind of escape plan from Earth has been theorised by NASA or some other organisation – whether some kind of exodus or departure plan is considered in the event of some future catastrophic event threatening our world. Of course, it would have to have threatened long enough for the plan to be implemented. Perhaps there is a pressurised module adapted to fit the space shuttle built with seats, recycling air systems and water purifying facilities. It would be well stocked with dried fruit and vegetables, miniaturised life support pills and protein wafers. A special power source bolted into place on the shuttle would propel the craft after leaving orbit for a long time, gradually building up its speed to an enormous velocity. Another unit would be used for braking and deceleration purposes. One wonders just who would be included in the passenger list and what professions and skills would appear be most required. Clearly, they would have little chance of utilising such skills *themselves* but could adequately tutor the offspring.

Would it possibly head for Proxima Centauri, perhaps simply hoping or assuming that at least one Earth-like planet was orbiting one of the stars in the system? Just what kind of people would be selected? If, for instance, you had to choose between a nuclear physicist and a mining engineer, you might think the nuclear physicist would be a clever chap and be handy to have along. But would he? Without his power station, what other function could he fulfil? Now, given that our intrepid band would have to start the whole human race over again from its beginnings (minus the Stone Age), the mining expert might be the chap to choose.

Most certainly a toxicologist to make sure the entire ship's company are not poisoned by the first attractive looking fruits they pick, and an agricultural expert would not go amiss. Clearly, it would have to be a very careful choice, as indeed would the books and manuals to take on the one-way trip – a very difficult choice. Data on the mining and production processes regarding metallic ores would seem to be essential for such a voyage.

If the ship finally arrived on a planet suitable in all respects except that it was populated by bloodthirsty creatures that would devour the crew without a second thought, the old human aggression would come in handy and weapons would have to be included on the ship's inventory. The old 'them or us' situation may again arise, and even though it would mean terminating life forms indigenous to the planet, clearly they would use such weapons if it meant the refugees' own survival,

Given the 'precious' nature of Earth-like worlds and their possible rarity, it is possible that very advanced alien beings that may be present in Earth's space, view us as some kind of planetary 'pestilence' with our treatment of the Earth and our leaving a horrifying legacy of radiation poisoning to our descendants. After the Chernobyl disaster, 6,000 people died and another 2.2 million people were affected by the fallout, and there has been a *92 per cent increase* in childhood cancer.

Great tracts of Belorussian land have been horrifyingly contaminated for a thousand years after the explosion of the nuclear power station there a decade ago. Highly contaminated graphite and uranium dust poured down on the land and its population delivering doses of up ninety times higher than that received by the victims of Hiroshima. Detrimental, if not lethal, effects will be experienced for many generations to come entirely through human ignorance, inefficiency and incompetence.

Russia cannot be blamed alone. Near-fatal accidents have been narrowly averted elsewhere, and it is a case of 'There, but for the grace of God, go we!' Russia does not have the immense wealth and resources of some other countries, and things are even more difficult today in its turbulent but brave transition to democracy, but to cut corners in financial budgeting where the safe containment of nuclear energy is at stake is suicidal. Other reactors still operating there are equally dangerous and lack concrete protection domes and water cooling jackets.

If we assume that highly advanced alien beings are observing all these Earthly events, it is reasonable to conjecture that their assumed advancement surely would not have been achieved without some records of failure, disaster and mistakes in their own history and memory banks. One could take the view that there would be more of an inclination among such beings to encourage, help and assist beings struggling to achieve high advancement such as ourselves, rather than viewing us as a planetary pestilence of some kind who are harming ourselves through ignorance.

However, by now, our own mental maturity has surely reached the point where we should be able to envisage the possible outcome our actions by using computer-enhanced scenarios. We are aware of what produces the oxygen so essential to our life forms, yet we continually destroy that source to make way for concrete. Sometimes we seem intent on concreting the entire land masses of the Earth.

When being educated, one of the first things we learn about the Earth is the drawing up of water from the Earth's seas and its formation into rain clouds, and these in turn rising up when encountering the mountain regions and precipitating the water on to the land, only for it to run back to the sea.

We also hear of plant life giving off oxygen and absorbing harmful gases like CO_2 which we breathe out. Now, the continual stripping of the trees and digging up of grasslands should in theory upset the balance, yet it doesn't seem to do so. The proportions of nitrogen, oxygen and other gasses seem to remain the same. Surely it should change. But then we hear of diatoms, the beautiful, precise, jewel-like symmetrical objects we find in sea water that may be responsible for keeping the balance. So we promptly set about them by dumping great quantities of oil and other pollution into our seas. The Earth is fighting back, but how long it can maintain the fight is another question.

It is quite alarming that as long ago as the Kon Tiki expedition, when Thor Hyerdahl crossed the Pacific on his flimsy raft, he was never at any stage out of sight of *some* evidence of man's pollution, even if it was just a few floating bottles and some shellfish stuck in a lump of congealed oil sludge. In the worst scenarios, great floating mats of the stuff are now encountered: oil, plastic containers and other rubbish all lumped together in a tarry mass. Suffice it to say that there have been many supertankers built since then, many disasters, leakages and purposeful tank flushings into the sea since those times. We are so used to seeing this junk floating in the sea nowadays and being washed up on the shores that we take it for granted.

Yet the dear old Earth puts up with it all and still ensures we have the same precise proportions of gases so essential to our life forms. Clearly, a stronger force than man's destructive attempts are at work. Do we have an ecological death wish? As said, there doesn't seem to be much point in thinking about the great things we plan to do, or where we may travel to or explore and what planets we will single out for terraforming and how we will go about it, or about bases on the moon and on Mars and deep space travel when our own immediate future here on Earth may be in serious doubt.

The Earth has been described as a living being. If it is, then we could be seen as a virus attacking it. One wonders what the antidote might be. If there are all these extraterrestrial vehicles flying about our skies, they would be in a good position to know how rare Earth-like planets are, and would be very confused and perplexed by our behaviour in our inflicting such wilful damage on our world.

As said before, it would all seem rather pointless to spend these huge sums on SETI programmes if the US government actually *did* have wreckage, or even the occupants, of these craft. If they do not have such evidence, the continuing stream of UFO reports will not go away, so why not try to solve it once and for all by diverting some or all of those funds into a prolonged sky search, using many fish-eye-lens cameras to watch a vast area of sky and log and record every movement across it? After eliminating aircraft and everything that moves upwards including birds, the unexplained objects reports will probably drop to almost zero.

Then we could begin to speculate and analyse those far-off but not too far-off star systems. It would be highly irritating and frustrating to discover an intelligent signal source from say 1,000 light years away, when *we* the discoverers would be long dead before they could be signalled or contacted with a return message. It would take 1,000 years just to say 'Receiving you loud and clear.'

Of course, it is perfectly possible that there may be civilisations 'out there', not so far away, that may not have the faintest clue about radio astronomy and may be just carrying on their lives in blissful ignorance of such things. But if that is as far as their intellect had taken them they would not be comparable to humans where such activity is merely a *stage* of development.

They may just quietly continue growing, tending, gathering and eating crops, and living and dying. They may be one with their planet, respecting it, only taking what they need and putting back all they can to ensure the continued enrichment of the soil. In any event, we will continue (since we have developed the technology) to search, listen and note, even though all we seem to receive is the gasping static sigh that is the remnant of the colossal event that started it all so long ago. We seem powerless against the desire to eventually leave Earth for the stars.

When listening to the whisperings of the universe, we have to scan many different 'lines'. We speculate on the likelihood of using the hydrogen line. It is logical to us and therefore should be to them. The hydrogen line (or radiated frequency) comes in at 1420.4 megacycles, and this was one of the first frequencies analysed.

On an Earth-like planet with a similar mineral composition, intelligent beings would experience everything we have experienced in the past at one stage or another. Fire, the actions of lightning and the discovery of metals. Their sky would no doubt produce a rainbow just like ours. Their natural intelligence would figure out how to reproduce it and what caused it. Once they started to analyse and think about light and other radiations and waves, they would be well on their way to their first astronomical radio dish.

Although we frequently fall into the trap of assigning humanoid-type qualities to hypothesised extraterrestrials, it really is quite impossible to speculate on their appearance and, given the fantastic diversity of life forms on this planet, anyone's guess will do.

However, if we are to give some credence to the many and widespread abduction claims, it would seem that a pattern *is* emerging with regard to the appearance of the alien beings.

Certain features prevail, such as large heads, large tapering black eyes, small or slight body frames, the absence of a nasal protuberance and a thin slit of a mouth. Rarely are they heard to converse. Instead the impression is always given of mental communication.

If one of our previous scenarios did happen to be true, if they are the other offspring of ancient astronauts and are coming here to see if we who share this ancestry have survived, then we would expect them to have at least a passing resemblance to us. After all, it is possible that we all looked pretty much the same at one time in the remote past. If they had discovered something profound that enlightened them about our world, they would be perplexed by us and our behaviour and not a little alarmed. They would not just aimlessly watch us day in day out, but would want to know why our anatomy had diversified and was undeveloped in areas where they had advanced.

They would be interested in our ancestry and evolution and what our common ancestors had been up to on this planet. They would be interested in examining the ancient landing areas and how we had fared in the dim past. On their own world, after their finding and analysing items and ancient texts from a time capsule, they may have discovered, they would be probably well aware how the refugees to their world had fared before degeneration. Their evolutionary studies would be clearly defined. They would probably have ample bone fossil finds showing a gradually changing anatomy, skull size, eyes and a gradual obsolescence and disappearance of teeth.

Their diet may well prove to be mainly vegetarian. Of course, their planet may be so similar to the world their ancestors departed that planetary forces would not have changed them greatly, except perhaps in skull size and a slighter frame. Their computations during their voyage to our system would probably have led them to expect us to have an anatomy and development similar to theirs, but more robust and muscular to handle the extra gravity and survival activity such as running and hurriedly climbing trees to escape predators and, of course, tribal battles and physical combat.

Before our hypothetical aliens began abducting live specimens they would have been able to obtain quite ample supplies of arms, legs, heads, torsos and any body parts they cared to analyse simply by descending on one of the many Earthly battlefields. One could imagine their laboratories on their polar or undersea bases with reactivated heads, their eyes watching and following the aliens' every movement as they lay propped up by their supports, being nourished by an ample blood supply and the appropriate electronic stimulation. Also limbs responding to stimulation, skin, blood, DNA profiles, central nervous systems, and bone structure etc. all being studied, analysed and logged as they learned more and more about our metabolism.

They would have to have a plan that would need to be clearly defined in stages, and this would be one of the early stages. Of course, they would be studying *all* the life forms of the planet along with its fauna, vegetation, and meteorological and lunar effects. They may have no moon and find ours very interesting with plenty to keep the occupants of the base there quite busy.

So what if 'they', our hypothetical aliens, *do* arrive after all this ground work and we rather surprisingly do accept them? (If we have a choice that is.) What would be some of their objectives? Complete re-forming of the military into a planetary protection group? First tasks to dig up and retrieve buried and sunken nuclear waste and, working closely with NASA, use the shuttle fleet to fire it into the sun? 'We will stand by in case of any accidents with your shuttle.' The aliens could probably take it all on just one of their starcraft, but they want to rub our noses in it a little bit and make us think we are cleaning up our own back yard. Eventually, they would get round to their primary objective.

Once all the signs of social disorientation and cultural shock had subsided and they felt they were being accepted, the aliens would begin their massive programme of first weeding out the faulty beings and removal of our more primitive and simian traits. The quasi-human specialists already installed at the top of all the branches of the medical profession begin drawing up their lists. Many humans will have their reproduction capabilities suspended. Everything is done in an orderly way, slowly and subtly and over a comparatively short period of time. Their aims would be to bring us forward in 'one giant evolutionary leap for mankind'. 'Einsteinean' intellects would begin to appear everywhere. Half of the class in any school would be child prodigies. Gradually we find the killing and eating of other creatures abhorrent. Our food processing would change and become more efficient and there would be no starving people. Large cooked meals would be a thing of the past. Just as scientists wrapped up in their work do not eat properly and do not even feel hungry, so most people will be wrapped up in intellectual pursuits and only occasionally stop to pop their vitamin pills and drink their vitaminised liquids.

What if the aliens' and humans' common ancestors did originate on Mars? One thing that would alarm the aliens would be our space programme and plans to go to and ultimately to colonise Mars. Before they could allow that to happen they would wish to make us completely aware of our identity and past history. They couldn't have us going there and digging, boring and blowing up areas for analysis, especially the area of the planet that was declared a 'shrine' and may still contain the fossilised bodies of those of their ancestors who did not escape the catastrophe there so long ago.

However, they may not wish to reveal themselves to us yet. They may have much to study and evaluate and, in any case, it will be some time yet before mankind leaves for Mars. If generations of alien beings have been observing us since biblical times and see no particular advance towards a peaceful co-existence – in fact they see us continually slaughtering each other – their conclusion may be to consider actively 'adjusting' our behaviour through neurological surgery or some kind of stimulation of the brain's unused areas.

During their supposed periodical meetings to discuss the 'human question', it would be bound to be commented on that during the supposed observation period of some 3,000 of our years, there had never been a time when war, or at least a battle, was not raging somewhere on the planet. Some being would be almost sure to make the comment that it was his fervent hope that they never meet on their celestial voyaging beings such as us with an equivalent or a more advanced technology.

One could imagine the dialogue at such a hypothetical meeting, with the presider opening, 'We are here, to discuss human behaviour patterns.'

An alien observer says, 'Their technological advancement in microelectronics, robotics, genetics, surgery and general science is advancing far ahead of their inclination to use such advances in peaceful ways, and strong negative forces still prevail in them. It is not our way to hold back the advancement of intelligence but to promote it, but there is a clear danger here. Our course of action is obvious: stimulation of their inactive neurons to bring their dormant processes into play to counteract their negative behaviour patterns and enhance their mental maturity on a par with their technological achievements.'

'Hear, hear.'

'Agreed.' and so on.

Another alien observer continues, 'Perhaps another teacher of righteousness, since the original plan was highly successful? In fact, Earthly doctrines prophesy or *expect* a return visit.'

The presider may answer in the alien equivalent of, 'They had their chance and they blew it! Our abduction operations are proceeding quite well. We are learning much. There was little point in direct mental contact in our early days of observation, but now they understand more of the topics we communicate on. I would like to hear your views on the suggestion that we should abduct Earth beings in much more influential and powerful positions and inject influential inputs into their senses. This would be important with regard to their military officials and would, in theory, cascade down to lower echelon groups, and the end result would be a changing influence in their behaviour patterns, perhaps avoiding the necessity for direct neurological action.'

Alien observer again, 'In view of our extensive observation processes, I would wholeheartedly recommend the first observer's comments on direct neurological activation. All our tests and invasive surgery so far carried out seem to indicate that the areas that would control or suppress their aggressive and war-like tendencies lie in the dormant area of their brains – the unused, inactive and unstimulated regions.'

To come back to reality, it's a bit frightening to think that this kind of dialogue could be going on above our heads, but it just might be. When despairing of our more negative and destructive activities, surely it's not all gloom and doom. Many people talk of high crime rates now and (although it would be no use telling this to a victim of a mugging) the figures show that the fear of crime is greater by far than the statistics.

In foggy Dickensian London, robbers and footpads lurked on every corner and all people worth robbing took hansom cabs. It seems it was largely the poorer ones who walked the streets. Also, murders were far more commonplace, as were pimps, prostitutes and social diseases.

Our alleged aliens, if they have been watching us for so long, must see a general trend towards or a striving for peace, non-aggression pacts, the League of Nations, the United Nations, UNICEF, Save the Children, Help the Aged, social welfare, venturing into disease-stricken and starving areas, five year plans, aid packages, birth control advice, worldwide immunisation programmes, agricultural science and education in Third World countries. Surely, it would be seen that all is not lost and not *only* negative and destructive actions prevail. There are as many positive thinking action men and women as there are negative, destructive, war-like and criminally minded beings.

'But that's the whole point,' the aliens could argue. 'There are *none* of the latter on our world. Why must you tolerate them here? They will always be with you. Your brains are locked with positive and negative aspects battling with each other, due to the lack of corrective surgery by our common ancestors who assumed such negative genes would be naturally overtaken and consumed. We now offer you that chance.' I wonder who would be the first to volunteer for surgery in such a 'nightmare' scenario.

As said, we are not now so destructive and savage. We no longer have Roman legions marching across our lands, Atilla the Hun sweeping across Europe or Norsemen raping and pillaging, so things have looked up a bit... haven't they?

Although we still have a long way to go before we can claim to be truly civilised, our mythical aliens must have been much more alarmed in 'olden times' when the bodies of crucified victims lined the Appian Way, for instance. Their main concern would be our *continued* appetite for war and the weapons to wage it, and how long it would take us to reach mental maturity, if ever, without their corrective manipulations.

It is interesting to speculate on what we may be capable of when and if we do use all our grey matter. We may have seen glimpses of it with the child prodigies and the geniuses of the world, and the glimpses into what we at the moment call the paranormal. We may, in the future, have to drop the 'para' from the word. Things like precognition, ESP, levitation, telekinesis and so forth.

As said, how many times has it occurred to people to say to each other, 'I knew you were going to say that', or they say the same thing simultaneously. Everybody does it. They cannot all be coincidences, and so our unused brain matter may be stirring itself at last. Perhaps we will eventually *all* communicate telepathically. The areas that may counteract our dangerous, negative and war-like aggression may be slowly developing and, as said, could be under patient and ongoing analysis. The original hypothetical 'experiment' may have been the introduction of a better design that resulted with Cro-Magnon man and his superior brain. In this case, the 'better design' would be our *increased* brain power, allowing us to continually upward instead of peaking and troughing like the cultures and empires of the past. What would the Greeks, Romans and especially the ancient Egyptians be now constructing if they had continued on the upward trend without the enslavement, butchering and wasteful conflicts? We can now gaze at buildings, built in a volcanically unstable area that have stood for over two millennia, yet we have frequent legal action today over tower blocks or bridges that have collapsed due to 'modern' building methods. We appear to have lost as much constructional 'know how' as we have gained.

In the section on creation we put the blame on the Almighty for the plagues, pestilences and corrective retributions, but we may equally blame our ancient astronauts for this action and assume *they* were responsible for attempting some form of crude action to ensure only the positive-thinking beings survive. As said, they would (if we assume they have been watching us since biblical times) now have another 2–3,000 years or more of advancement under their belts and, whereas their craft may once have roared and flashed and gone forth on pillars of fire, their power sources and craft would be that much more advanced and allow them to perform the marvellous feats of defying inertial and gravitational forces which we attribute to them today.

With human beings seemingly the exception to the rule, it does seem logical that the general advance of technology should bring about an equal advance in positive and civilised behaviour patterns. But given that our alleged ancient astronauts were responsible for the aforementioned retributions such as Sodom and Gomorrah, turning people to pillars of salt and drowning entire populations, and for the Ark of the Covenant which seemed like a horrific secret weapon, the equation of technology equals civilised mental progress seems to be just a typically human evaluation. 'They' may not be so benign and peace-loving after all, but quite savage when it suits them.

If we do survive all the hazards, both celestial and those of our own making, and if the alleged aliens do not interfere, then we have this vast span of time ahead of us that is the remaining life of our sun which should amount to another 4.5 billion years. 4.5 billion years of technology ahead of us – the mind truly boggles!

In theory, by then we should know everything, all the secrets of the tiniest particles, their activity and actions. Quantum physics will seem like our present day ABC or times tables. Our bodies will have long been replaced by God knows what. We really would not know ourselves. We should be able to streak between the galaxies, let alone between stars. Many suns have lived their lives and died. Where are all these people? We are bound to ask again – are we the first? Was divine creation right? Are we expected to 'go forth and multiply' on a much grander scale than mere earthbound parameters?

We certainly will not be hanging around this neck of the woods when the time comes near for our sun's demise. Maybe we will live in the Belt of Orion or somewhere further away. Far into the future, which will become other beings' past, there may be planets between here and that faraway place with budding technologies examining *their* past histories and finding the basic algae in their primitive soil that *we* put there in passing; an action that ultimately produced *them* and caused *them* to query and wonder about it all.

If an extraterrestrial intelligence did enter our earth space (or already has) and had reached the hypothetical state we envisage possible for ourselves, it is doubtful we would even be aware of their existence. It does not seem likely that flying saucers would be their mode of transport. We may logically suspect this kind of advanced intelligence has not yet been reached by anyone and that the head gardener had come along and put a stop to the pilfering of all those apples from the •tree of knowledge long before anyone got to the last few.

We like to feel here on Earth that we have reached a high degree of intellect, yet anything out of the ordinary is left to a few dedicated enquirers to solve, is shunned by science and assigned labels such as 'fringe' or 'paranormal', simply because it won't slot into those established patterns. Yet one would think the opposite would prevail and that these things we flippantly label as paranormal or lunatic fringe activity would be the first things this questioning, enquiring mind of ours would wish to solve and probe into, if not for sheer curiosity, then as a challenge.

Of course, scientists have to eat too. Research institutes would not get grants if they called themselves 'Ghostbusters Inc.', or 'Alien Trappers Anonymous'. It would be all very well if one were a retired millionaire and scientist (and there are not many of them around) to finance and carry out one's own research.

Of course, as long as these odd topics are left unsolved and pushed into the corner, the longer people will have fun speculating about them and would surely miss them if they were gone.

The area in remote Siberia of the so called 'Tunguska event' of 1908, when a crater-less blast or 'airburst' killed herds of deer and levelled trees for many miles around, was not visited for a long time afterwards because of this remoteness.[10] It seems there was less scientific curiosity in those days and until the advent of flying craft and their operation in the area, the obvious ground effects were unappreciated. The current thinking is that it was caused by a comet. Plausible enough of course, but perhaps the Russians have the real answer themselves. The explosion was apparently heard 750 miles away and reverberated around the Earth a couple of times. It was compared in its explosive violence to a 30 megaton nuclear device, *yet left no appreciable crater*. Between 7–800 square miles of tundra were affected in the ensuing blast. Because of this aforementioned remoteness, news of the event, which of course the local peasants were well aware of, did not reach the Western world until twelve years later. Although no human casualties were reported, it is almost certain that there were some as people were thrown off horseback forty miles from the epicentre.

Was it just fortunate coincidence that this spectacular event occurred in such a remote area?

If its approach course had been varied by a couple of degrees it would have obliterated London. Hints of it being a stricken spaceship are based on this fortunate coincidence of being 'steered clear', so to speak. Pilots of stricken aircraft quite often steer their craft away from built-up populated areas. Apart from the heroics involved, it is a lot more comfortable to put down your craft in a grass field than on a row of houses.

---

[10] Jack Stonely, *'Tunguska' Cauldron of Hell,* Star, 1977.

If it had been formed by a comet and got that close to the ground, there would have been enough material left in the comet's head to have produced at least a modest crater. If it is still possible to investigate it after all this time, perhaps modern instruments at our disposal may solve the mystery. However, the area is quite hostile, bone-numbingly cold in winter and a mosquito-ridden bog in the thaw. If we ask, could it happen again, the answer must be yes. If we ask, have we any protection from such things *nearly ninety years afterwards*, the answer must be no. We have the *ability* to protect the earth but not the co-operative will.

The event was investigated in the twenties by a certain mineralogist called Leonid Kulik (the poor chap later died in a concentration camp just for being a Russian Jew). Some of the peasants in those parts are given to longevity. Whether this is due to the clean, sterile air or not is open to question, but even now there would be eyewitnesses of the event still alive. Perhaps it would be a good idea to have another talk with them before they 'pop their clogs', because there seems to be some contradictions in the account and some doubt whether there was a single steady course of entry and an arcing across the sky to impact with the ground zero, or whether it zigzagged, turned, changed course, or whatever. If it turned out to be the latter the implications are quite serious and the sooner we know, the better.

At the same time, perhaps we could clear up the point about the silvery particles supposedly found in abundance in the local mosses at Tunguska and not found anywhere else on Earth.

It is not surprising that rumour, conjecture and speculation rule when we have such an absence of determined scientific investigation and conclusions. If such an object did strike a major populated area, former considerations of expense and suspicion by the military of the misuse of orbiting devices would indeed seem petty with hindsight after such a catastrophic event.

I cannot remember the source, but I seem to remember reading somewhere that quite a large asteroid actually passed at one time in our recent history between the Earth and the moon. That is frighteningly close and it would not be surprising if such a thing (providing it was not leaked) was kept from us. However, there are a lot of amateur astronomers about these days and one cannot imagine them all keeping quiet about such an event. But then they might not all be looking, might they? To be sure, a lot of people would be killed or injured in the ensuing panic that would surely follow if the residents of New York were told, 'Keep your heads down, just in case it does land on this city',.

More frighteningly, it has been said that asteroids have sometimes been detected moving *away* from the Earth and were missed or undetected in their *approach.* There is ample evidence that these things have hit us in the past, not only by the obvious evidence in Arizona, but with the advent of satellite surveys revealing craters of some considerable age. It is computed that these events can happen every 50 million years or so, which makes us uncomfortably nearer to the next than we are from the last. In the seventies, a smaller one came by at a distance close enough to 'brush' the atmosphere. (It is okay to tell us now!) In cosmic terms we could have reached out and touched it and, of course, there are plenty more where that came from. And they wonder why there is not enough material in the Asteroid Belt to form a planet! When our satellites reveal eroded meteoric craters on Earth, how many impacts have there been in the sea (which covers 70 per cent of the earth's surface) which have caused tsunamis?

The theory that the Tunguska event was a stricken spacecraft may not be so far-fetched if we allow that these oft-reported objects we perceive flashing about the skies are, in some cases, extraterrestrial. Surely the sixty plus different shapes and sizes that are reputed to fly about cannot all remain serviceable. Surely the odd fault must develop from time to time, especially between 100 billion mile services, and especially as it would appear that a lightning bolt can bring one down in the New Mexican Desert and scatter its occupants about the place. In spite of all their technology, it appears that they are still as vulnerable as we are.

A very interesting theory as to what powers the alien craft possessed was attributed to Kenneth W. Behrendt, and was published in *Phenomenon* by John Spencer and Hilary Evans (Futura Publications). They spoke of the 'anti-mass field' theory which is that an object could have its gravitational and inertial properties cancelled out, enabling it to defy those forces in its flight.

In the assumption that a body has a non-electromagnetic 'mass field', which one may suppose anything with mass has, then the imposition on it of an 'anti-mass field' of equally strong proportions would, in theory, result in an object with no mass field. This, coupled with a suitable driving force, could enable it to carry out the reported and seemingly impossible manoeuvres of high speed changes of direction within our atmosphere since they would be freed from drag and also inertial and gravitational restraints.

One assumes the theory came about due to the often reported rotating of the craft or at least of the rim of the craft or saucer, and this in itself implies some sort of power generation. In the case of this assumed power source and 'anti-mass field generation', it is assumed that the UFOs that darted about from the earthly 'foo fighters' of the Second World War, down to the objects that wound up on the tails of pursuing supersonic jets, were all equipped with anti-mass field generators. Of course, coupled with this anti-mass field capability they would require a driving force and this could be achieved by a 'plasma dynamic drive' which, loosely speaking, converts the air surrounding the hull of the craft into plasma. After a certain process, this flows around the hull without touching it and then pushes against the stationary or 'untreated' air at the trailing edge of the craft. The well-known Newtonian law about equal and opposite reactions then applies.

Whatever propulsion systems the alleged ancient astronauts were employing in biblical times (and their descriptions sound quite primitive), they have had over 2,000 years of modifications since then. Since they could bridge interstellar distances *then*, their power sources *now* must be quite formidable and, of course, other branches of their technology would theoretically be equally advanced. No more blinding lights and voices booming out from mountain areas, burning bushes and pillars of fire! They may have frightened the ancients that way but not us. Oh no! We take a bit more convincing. It seems one of these craft would have to land in our back yard before we would believe. Maybe not even then, unless we could look inside.

While not detracting from man's capabilities and obvious brain power, we do not know for sure if we have ever invented anything for ourselves.

Given the mind scan and telepathic techniques used in the alleged abduction cases, where all fear is removed and humans are putty in the hands of the aliens, we could have been subconsciously directed towards the achievements we give ourselves credit for from the first extraction of metallic ores to our latest inventions. (However, this means that our hypothetical 'creators' would surely have to take some responsibility for our more negative behavioural traits.)

All this has a specific purpose of bringing us ever closer to the point when 'they' may risk open contact with us, or be able to communicate, if not on equal terms, then at least in an easier fashion. Just as they think they might be closing the gap a bit and getting nearer to this point, we go off to war again somewhere on the planet and perhaps put back the possibilities of contact another ten years or so.

4. The Ultimate Purpose?

With regard to the sheer volume of UFO-related phenomena, alleged abductions and so forth, responsible governments would not be considered responsible if they did not have at least some loosely formulated plans in case these possible extraterrestrial theories became fact. Even if there are no slim-bodied aliens floating in tubs at such-and-such-an-airbase under wraps, one can quite easily imagine a red file in MOD or the Pentagon entitled 'Extraterrestrials – Actions and Options' that is brought out and dusted off at the periodic UFO 'flap'. Open the file:

> *Contingency Plan A.*
>
> *In the event of extraterrestrial intelligence manifesting itself to the point of belief and not conjecture, and/or confirming their presence by selective contact to high offices of this country, the following list of military and scientific departments and their current heads*

> *will come into play and be immediately informed and activated...*

There is a mass of information to correlate, compute, analyses sift and conjecture upon and if at the end of all the computation and analysis we still have a negative answer, then we never will have one until 'they' choose to give it to us.

It is hard to think of any description of a flying craft, method of contact, type of propulsion or alien body shape that has not already been seen by somebody somewhere on Earth. One thing is for sure: we cannot change our make-up or ways of thought and actions at the drop of a hat. We are what we are for good or bad and it is certain that if an extraterrestrial intelligence did come into our air space, no matter whether it was for our own good and what plans they had for us, there would be the immediate formation of a World Resistance Movement. If people are around today that are willing to drive vehicles full of explosives into installations of a hated enemy or regime, with what we could term the 'Kamikaze syndrome', they would emerge then. We would find it hard to resist the natural human trait of seeing the aliens as an enemy.

Also it seems that any tight situation, tension or war between countries seems to produce the 'right man for the job'. The Eisenhowers and Montgomerys and Pattons and Churchills all come to the fore. Now it might be: 'We will fight them in space, we will fight them in orbit and on the landing grounds, we will never surrender'. It is not in the human spirit to give up easily or without a fight. There must be (so the thinking would go) some weakness, some chink in their armour, some vulnerable point, perhaps a biological one. After all, the whole human psyche centres around adversity. We 'fight' for survival, we 'battle' against the elements, we 'conquer' disease, and so forth.

Given the end results on Earth in the past of more advanced races contacting less advanced people and the latter putting their trust in the former, it is almost always to the detriment of the latter. So it would be with extraterrestrial contact. No matter how much they reassured us that their appearance would ultimately be beneficial to us, we would assume (fairly naturally, one might suppose), that their ultimate objective would be the subjugation, domination or at least full control of the Earth and we would react as formerly stated. If the extraterrestrials have been watching us for as long as some allege, then they should know this fact. So where does this leave the somewhat irritated and frustrated extraterrestrials?

Or are these just Earthly emotions? Perhaps the aliens are infinitely patient and benign. If so, they will theoretically keep watching us patiently until our behaviour patterns moderate and in the meantime slowing or preventing our interplanetary adventures, except those to and from our moon. This may have happened with all those rocket explosions and losses before the Apollo programme, the loss of the 1988 Russian Mars probe 'Phobos' and the 1996 probe that did not even leave (or even achieve) Earth orbit, but fell into the Pacific.

If they are not infinitely patient and benign, then once they have shown their hand by open contact with us, it seems they would have three main options open to them:

1. To continue dialogue with us from orbit until we see sense and agree that we need their 'guidance'.
2. To land and enforce (mentally) their will on us and tolerate, or deal, with the 'Earth Resistance Movement'.

3. To go away and never speak to us again.

Option three can be ruled out immediately, as through long and patient observation they have seen what we are capable of. They know of our Mars voyage plans and more ambitious 'Project Daedulus' and 'Project Orion' and have noted our 'star wars' developments and theories about interstellar travel. We are down to two options and now, through simple logic, we will eliminate option one. This would require them to be infinitely benign and patient. We would never willingly agree to have our minds bent and, unless they were 'god-like' spiritual beings which they are not, they would not be possessed of an infinitely benign and patient mental disposition. Any exploring, creating and questioning being is always anxious to see the results of those actions and, being anxious implies impatience. Therefore, they would expect a result or culmination of any plan or action within a reasonable time, which clearly, only leaves option two.

Perhaps then we will compromise in good old Earthly fashion. 'They' will make contact and impart to us certain information; we will digest it over a period of time and come to a decision.

Perhaps this information is so profound when imparted to us in all its detail that we will openly accept their guidance after seeing ourselves for what we really are. They may use a form of brainwashing by bombarding us with explicit scenes, facts and figures of human slaughter from around the time of Alexander the Great, and instead of scenes of us pressing buttons and killing masses of people, we will be treated to explicit full colour pictures of the Roman campaigns in their goriest battles, Atilla the Hun and the Viking slaughters, right on down to Hitler's handiwork. By then we might be so shell-shocked and reeling from all the butchery that we will submit to any guidance.

They may tell us that the 'human' type of being has been in existence since before our planetary system cooled down, that we seeded Venus, occupied it, ruined it, seeded Earth (during our ravaging of Venus to the point of a runaway greenhouse effect), moved to Mars, finally succeeded in blowing up Planet X (now the asteroid belt) and through this action inadvertently ecologically destroyed Mars, took up residence on the now fully terraformed Earth, and finally are now on our way to ravaging the Earth. But since we have taken no action to reactivate Venus, we have nowhere else to go.

So, after watching all these explicit videos of real people being torn apart in the Roman arenas and real blood and body parts flying about, we realise what we are and have been capable of in real terms, instead of the enjoyable Hollywood epics.

The extraterrestrials then (showing off a bit) show us Earthlings their wondrous technology: a massive mothership lands near an area of high metallic ore concentration we did not know about and is now shown to us by the kindness of their hearts. The massive doors slide open and a huge machine trundles out, bores, extracts a mass of ore, smelts, forges and shapes it. Gradually a framework appears before our eyes; sheets appear like newspapers off a press and are inserted into place. The machine trundles into the massive hangar-like building and takes a bow, after completing its building feat in record time.

Now we are aboard one of their craft in space, the gleaming white hold suddenly interrupted by a thin black crescent that becomes a circle. It is full of stars. We realise a hatch has slid open and now we are moving towards it as though we were inside a tube lying there, looking out of its glass front. We are streaking towards a rotating spitting mass of a cometary head. We can hear the particles striking the device as we plunge into its heart. A brilliant light, then darkness. We have just been treated to a weapon's eye view of a planet-threatening comet being destroyed as it approached their world in a very threatening manner.

They treat us to a view of a hostile-looking world with three suns in the sky, one a large bright large star. They point out our star to us – just a speck of light – and inform us they knew of our sun and planets before our common ancestry with them was known. They show us an early 'planetary council' meeting on their world discussing humans and our behaviour patterns. The ruler is communicating to his people:

'They are thousands of years behind us. The humans are barbaric and warlike. Contact has been ruled out for the foreseeable future by our observers in that system. We must formulate a plan with a view to modifying their behaviour patterns.

It requires a teacher of righteousness created in their image for close contact to impress them initially, then gather a following. There will be some hostility, perhaps even violence, but short of destroying him by fire ('Though he die, yet shall he live') he can survive any of their other barbaric forms of death.

We are possibly treated to the base being constructed on the moon, close-up views of Earth and our achievements, even our astronauts capering about on the moon. Eventually they hope that we will be in a suitable enough condition to throw up our hands and say, 'Okay, you win.

What do you want us to do?' That is how they may hope we would react; in reality, our reaction may be quite different.

It is a fact that many people disappear without trace every year, as do many aircraft and ships. We read sinister implications into this but, as a certain professor once said, why don't trains disappear? It implies of course that the natural hazards confronting ships and planes ensure a good percentage of them would disappear. There are strange meteorological anomalies at work in the air and at sea that we do not know everything about.

I can recall an incident in foul weather in the Middle East when an aircraft of our squadron had to be scrambled to 'shepherd in' a stricken civil airliner that had lost airspeed indication, radar and other instrumentation in a freak storm. On landing it was noticed that the fibreglass nose cone containing the radar had been battered off by hailstones the size of golf balls, that the pilot said were flying *up* at him as well as from other directions. It was a tribute to Rolls Royce that his engines did not flame out as the intakes, or engine nacelles, were dimpled with the same golf ball sized depressions.

Even without hypothetical aliens, Earthly conditions can cause the loss of ships and planes. Nevertheless, some strange encounters are on record of aircraft disappearances accompanied by the frightened voice of the pilot saying he is being covered by some dark shape above. Then all communication is lost and no trace is found of the aircraft. One such case occurred in Australia involving a light aircraft that followed this pattern. So what do we make of those cases?

Perhaps the aliens, as part of their plan, instruct some of their number about all things related to Earth. They are then possibly inserted into Earth life with high objectives in mind, programmed to climb high in their professions in time for the 'Second Coming' or 'final plan'. Are these the MIBs, or what ufologists call 'Men in Black'?

There may also be many humans walking on some alien world in wide-eyed wonderment. Of course, the majority of missing humans must be due to accidents, unsolved crimes, drownings, people being washed out to sea by freak waves, falling down old mine shafts and many other reasons. If we do have alien beings watching us from the skies they could quite easily take people who would not necessarily be missed, but they would hardly get the 'pick of the crop', so to speak, being restricted largely to travellers and tramps.

Of course, with their alleged mind-bending powers employed in all these abductions, one supposes that with a bit of luck they could take pretty well who they liked. Naturally, it would have been even easier in earlier times before Earthly radar could detect their surface 'forays', and in biblical times any desert wanderer could have been collected.

One would think that with all the people they must have taken up for examination over all this time they would be heartily fed up with poking about with the human anatomy, but if they are looking for signs of genetic change and mental improvement, those few extra neurons now active since the last abduction programme would be looked for in each successive generation. Perhaps they are looking for signs that we are growing away from our more gross behaviour patterns by a lengthy and ongoing analysis. As by far the largest portion of alleged abductees are female, it would appear that collection of eggs is the primary objective. Indeed, in some hypnotic regression cases this has been the stated aim of the invasive probings. They may be producing an entire army of quasi-human beings.

They would 'have ways' of ensuring their installed part-alien beings obtained positions of high authority. Having such powers of telepathic mental control, the 'creators' could visit a person in their sleep who was known to be taking interviews the following day for a position of high authority and influence, and implant in his mind a certain instruction. For example, 'When the candidate with the red handkerchief in his top pocket arrives before you, he will be deemed suitable in all respects'. And so the following day, the hybrid being during his interview removes his handkerchief and dabs his nose. Then the appropriate remarks are made on his sheet and he moves one more step up the ladder.

Perhaps when all the quasi-humans are in place, the next phase begins. A subtle form of brainwashing of all those around them gets the subject on to extraterrestrials, and their line might be 'If they have the technology to travel interstellar distances with ease there would be no point in resisting it,' or, 'We could learn a lot from them; why bother in pointless resistance? They may solve our energy problem at a stroke with clean hydrogen helium fusion, introduce unthought-of agricultural methods, and so forth.'

The alien beings, having overseen us for so long, may stay and ensure the progress of it all is periodically recorded on small discs and placed in something that looks like a microwave oven that converts the information into tachyons which streak off at many times the speed of light and are in no time inserted into the memory banks under 'Planet Earth Conditioning – Phase Four' on their home planet.

Of course our mythical alien types thus far, in spite of their alleged wondrous capabilities, are still humanoid. We find it difficult to imagine any other description and we always assign a humanoid appearance to them. Of course, it is entirely possible that advanced beings, encountering us may no more think of contacting us than we would of watching intelligent creatures such as bees manufacturing their combs, or termites building their clever air-conditioned dwellings. Perhaps they would just watch us with a passing interest. We watch and are amused by the antics of the creatures in the zoo performing clever tricks, but we do not go up and try to strike up a conversation with them.

Perhaps a clever trick with which we may have amused some vast cool intellect passing by was our astronauts bounding about on our moon. However, we must keep things in perspective. Compared to some other Earth creatures *we* are the vast, cool intellects. As previously said, perhaps one reason why we may not have been visited is that 'they' would rather go upmarket than downmarket, and they would be more likely to find more intelligent life on the older, inner systems than any that may exist in the outer galactical arm such as ourselves. Lesser mortals or not, we are still the prime species on Earth with our dominion over all others.

It is uncomfortable to dwell on such hypothetical beings of vast intelligence and intellect with unfathomable wisdom, because we are unable to draw any conclusions about how to deal with them. It is much more comforting to bask in our own fortunate over-endowment of intellect from such a comparatively short evolutionary period, which may be as short a time as 50,000 years. It is only natural that when we visit our zoos some people are affronted, religious or not, by a comparison to the apes as, although they amuse us, they show no more human-like traits than a dolphin. As predecessors of the human race the one most obvious failing when looking for human-like comparisons is their complete and utter lack of creativity.

The chimpanzee, allegedly our closest relative in the primates, looks docile and dull yet does possess intelligence of a sort. But when we start to look for attributes to compare them to humans, we begin to struggle. For instance, humans are perfectly at home in water. Babies born in water have their trauma of emerging into the world somewhat eased, and all babies seem at ease in water. If a chimp falls into water it quickly drowns: there is no affinity, desire or urge to come to terms with water or to learn to swim. Then there is the matter of body hair which we do not have at birth and they do. It is as though we are conditioned by evolution for birth in a much warmer zone or planet where this bodily protection from the elements is not required. Then there is the fast-growing hair on our heads and chins. No one sees a barber calling at the zoo to give the chimps their occasional trim. Many more important dissimilarities exist than any supposed similarities.

Extraterrestrials in the abduction cases are almost always described as smooth of skin. If we *are* a kind of hybrid, then we would have this 'half and half' attribute. The hair from the simian part and the smoothness elsewhere as a result of the alien part of our makeup.

The late Dr J. Allen Hynek was undoubtedly the guru of ufologists everywhere, and it was he who coined the phrase, 'close encounters of the first, second and third kind'. The latter, of course, formed the title of the famous film in which he himself appeared towards the end. The film dealt with visual and physical contact and the events depicted, apart from the final appearance of the aliens of course, were taken from alleged real life encounters.

Although the creatures that emerged from the craft at the end of the film appeared smiling, benign and peaceful, it is doubtful that the hysterical mother who had her child snatched or the parents and loved ones of the missing flyers who had never had their deaths confirmed saw them that way. Of course, the child was quickly returned to its mother, but although the time dilation effects may have made it seem to the aliens only a short time that the fliers were abducted, they must have been aware that a long time would pass for the flyers' relatives back on Earth. However at this stage it is not a question of whether they are friendly or unfriendly, but whether they exist in the first place.

The figures quoted for UFO sightings around the world are quite staggering, but a common misconception is that UFO means 'alien'. Of course this is not the case. It merely means 'not identified'. If when I saw my metallic circular object silently pass above me I had been in my back garden instead of being on an RAF base near the meteorological office that released it, I would have been convinced and ready to swear that I had seen a definite flying disc and, since we have no such craft (as far as I know), my conclusion may have been 'extraterrestrial'. When I say 'as far as I know' with regard to flying discs, I remember a magazine called *Flying Review* (not *Flying Saucer Review*) depicting a circular craft developed by the now defunct AV Roe Company. Apparently, it was a flop. The USA also experimented in this field and may have (it has been suggested) developed an actual flying craft which would have been viewed most certainly as 'extraterrestrial' if observed during secret trials. One example of this is the

Stealth Bomber which, because of its strange shape and high security, could easily have produced a Roswell incident, had it crashed during trials.

The aforementioned film appeared on the scene at the height of one of the so-called waves or 'flaps' that occur from time to time, which was an exciting time for the ufologists and a busy time for the debunkers. It is quite natural for the defence forces to attempt to 'debunk' them. As said, they would not want to appear helpless in the defence of their country from a possible unidentified threat, but whether the Condon report finally got the hot potato out of their lap and into the scientific realm is doubtful, as science also seemed to wash its hands of the whole affair. It would seem, even though they tried their best to distance themselves from the more highly publicised reports, that they still find themselves having to trot out the now familiar reasons for the sightings with the added item of 'space junk returning into our atmosphere'. Not really so strange when one considers the amount of stuff rattling around up there. There are said to be around 8,500 manmade objects currently circling the Earth. Prime time television slots in America probably quite rarely nowadays devote much time to the phenomena, but in the seventies they were certainly hot news.

The programme would attempt to get a balanced view, a ufologist or two and a Doctor This or Professor That who were usually the debunkers. There may have been present a victim of an alleged abduction, or a witness of a close encounter, and maybe even a retired astronaut (astronauts are always being said to have observed UFOs, but cannot be pinned down in most cases to say so themselves in public). The retired astronaut may have been more inclined to open up on the subject than a serving one who may find himself quickly off the programme for discussing such things in public.

The debates would sometimes become quite heated, the ufologists stating their cases and making allegations of cover-ups. The debunkers would deny this and state there was nothing to cover up, then trot out the usual reasons for the sightings, i.e. swamp gas, balloons, meteors, the planet Venus, ball lightning etc.

A certain professor who appeared on one of these programmes with a long list of credentials to his name was known to have stated that 'It is possible that Earth has been visited many times in the past by extraterrestrials', proceeded to ridicule on one of these programmes an alleged abductee who was made to feel like the village idiot. The professor, among his other achievements, was a recognised expert in exobiology or life forms off Earth – in other words, extraterrestrial intelligence. How you can be an expert on something you do not even know exists would have to be explained very carefully to many people. We do not even know for sure, in spite of all our technology and astronomical research, that planets even exist around other stars, although it is extremely unlikely that they do not. However, we *do* know that a planet exists containing a multicellular life form 23,000 light years from the galactic centre. It is called Earth.

One of the books I studied on the UFO phenomena was written in that busy time of the seventies for the ufologists and contained many of the classic sightings of the period. During a discussion in the book of the pros and cons of interstellar travel, as opposed to interplanetary travel, a certain gentleman by the name of Dr Robert E. Enzmann of the Raytheon Corporation, USA, was referred to as an 'expert' on starflight (or interstellar travel). Now, starflight does not exist today and it certainly did not exist in the seventies. So we have another 'expert' on something that does not exist, or a science without a subject.

One can imagine the scenario warming up as the accusations and denials flow back and forth, the guests having probably already warmed up a bit with some verbal jousting in the television studio. Our hypothetical guests might be a Professor E.G. Head from the American Institute for the Advancement of Fine Things, a well known ufologist Mr I. Sorritt, and an alleged victim of an abduction Ms Anne D. Tooker. The professor is speaking: 'I'm sorry, what plain?'

The ufologist replies, 'Nasca, the Nasca Plain in Peru.'

The professor looks at the audience, raises his eyes to Heaven and replies dryly, 'Oh... that one?'

A nervous laugh from the audience follows.

The professor pitches in, 'Yes... I've read those accounts. Quite frankly, I think you people are always far too ready to attach extraterrestrial influence to so many things without stopping to think of a more down-to-earth possibility. Furthermore, you detract from man's capabilities by assigning everything that seems a bit clever to the action of interplanetary beings. Really now – runways and aircraft parking bays – surely you wouldn't expect beings capable of interstellar travel to come into our atmosphere and then lower landing gear, or come down like one of our Earthly aircraft?'

The ufologist replies, 'I never said they were runways and parking bays, and neither did anyone else, only that they looked like them from the air. There are monkeys, insects, spiders, all kinds of creatures depicted that can only be seen from the air. We simply maintain that if extraterrestrials did land there and were observed analysing

Earth creatures, and maybe marking out lines for their Earth charts and maybe some kind of triangulation references Then the natives, anxious for what they perceived as sky 'gods' to return, reproduced the markings on a grand scale, ensuring they could be seen from the air. My other point is, that when you talk of a spacecraft coming into our atmosphere and landing, isn't that just what our space shuttle does? That's a spaceship isn't it? And it could be an interstellar one with the right power source, and it could land on another world with an atmosphere?'

The abductee finally gets a word in. 'You all talk so hypothetically. They are here, it's happening. How would you like to be in my shoes? I've had nightmares ever since the...'

The professor cuts in, 'With respect, are you really sure that it wasn't just that – a nightmare? Can you be perfectly sure in your own mind that you didn't just dream it all? After all, disappearing out of the window with them like Peter Pan and going up to a flying saucer?'

The abductee: 'Listen, can't you understand. It's all on the tapes. I...'

The professor holds up his hand and his voice takes on an overly patient tone. 'I know about the hypnosis sessions, but hypnotic regression only reveals what people themselves *believe* to be true.'

The other ufologist adds, 'Goddam it... excuse my French ma'am... there were radar and visual sightings across three states on the night in question. State troopers, police, ordinary citizens, all saw something moving towards her area, and the last report was around the time her watch stopped and the power failure occurred.'

The professor replies, 'I'm not denying this lady's belief in what she thinks happened to her, but many people see things – goblins, leprechauns and fairies. They believe they exist, but that doesn't make it a fact.'

The ufologist: 'What does it take for you people to believe?'

The professor replies, 'Do you have an artefact, something real other than a blurred photograph, something not of this Earth? Don't talk to me of 'angel hair' depressions in the ground, burnt leaves, corn swirls. There are plenty of Earthly events that can cause such things. You people deride the explanations given. Have you heard of the high altitude air streams? Airline pilots use them. They can throttle back and save a lot of fuel if they are going that way. If a weather balloon hits one, particularly the old silver painted ones, they can pick up quite a speed and offer a good radar return if there doesn't happen to be an aircraft in the area. Bingo! There's your UFO. Have you ever watched a bright planet low on the horizon if you're on the freeway and you happen to glance at it? It looks just like its keeping pace with your car, so you stop and it has stopped too. You think it's interested in you when it hasn't moved in the first place.'

Ufologist: 'When you say produce some material evidence, are you sure this government couldn't show us some material evidence? New Mexico, 1947 – am I getting warmer?'

The professor shifts in his seat a bit, 'I... wouldn't really know. I don't give those stories any credence whatsoever. Even if I did, my security clearance would not be high enough for a 'need to know'.'

The ufologist keeps up the pressure: 'It is fairly well documented that the military got out there pretty damn quick, cordoned everything off and got everyone out of the area, but not before a very credible witness saw wreckage and small humanoid beings like nothing that was of this Earth.'

The professor, who by now regained his composure, replies dryly, 'Hmmmm... interesting... Can we talk to him?'

The ufologist somewhat dejectedly replies, 'No, he's dead.'

The professor turns to the audience and replies with one word, 'Oh!' then moves into the attack. 'Look, what else would you have expected the military to do? There are experimental airfields out there, testing grounds, activities essential to our defence. Crashes have occurred there, sure. Its quite natural there would be a quick reaction to a crash in order to keep rubberneckers, ghouls, souvenir hunters and so on out of the area. It would be classified material lying about the desert there. Look, I don't disbelieve the possibility of extraterrestrials. My job isn't to deny their existence. I'm quite sure there are life forms on other worlds, but if they have interstellar capability, they surely would have radio signalling capability.'

The ufologist: 'Maybe we weren't listening at the time. The evidence is they've been here a long time.'

The professor: 'That's possible, but we're listening now, and so far nothing. No one seems to be talking, no one.'

The lady abductee gets another chance to interject. 'I thought you were sold on the idea of extraterrestrial visitations to Earth in the past?'

The professor: 'I've said that it's possible that they have, perhaps many times, but that doesn't mean they are still here.

There may be many things in the cosmos of interest to them besides us. We mustn't set ourselves up as something special. If they did come here, it may have been just another populated world to them and they simply noted, logged and examined and went on their way, and simply inserted us into the galactic catalogue under, perhaps, 'Dangerous Species'. Quite frankly, I don't believe we ever achieve anything in these debates. We just go over and over the same ground and, right now I have better things to do.' He moves to disconnect his communication device, then the host moves in to wind up the show.

The Condon committee, in the bulky book that was produced of the report, summed all the UFO reports up by saying it was doubted that they would advance the cause of science by further study and the reports did not appear to constitute a threat to the defence of the country. One does wonder if the word 'saucer' was not psychological suggestion by Kenneth Arnold in 1947. Anyway he did not say they were saucer-shaped at all, only that they flew like a saucer skimmed across a pond. The shape he ascribed to them was a crescent, which might suggest high flying geese, and the skipping movement the beating of their wings in unison. One wonders whether if he had said they were box shaped everyone would be seeing flying boxes.

Perhaps the cause of science *has* been advanced by all the conjecture, deep thought and pondering as to the nature of the extraterrestrial craft's power sources, as previously conjectured to be anti-mass field generators. What was happening during the so-called Philadelphia experiment? Experiments to become invisible to radar detection were supposed to have happened using navy ships and their crews in some sort of 'live' experiments, supposedly to produce a radar detection 'blanket'?

Apparently, there was an interesting situation in a French factory, when the cause of so many workers reporting sick was located to low frequency sound resonance. When science looked into and reproduced it on a large scale, the sound caused a brick wall to crack up. Shades of the walls of Jericho story. All the different things science has reportedly experimented with may have resulted in something now 'under wraps', and may all be due entirely to the surfacing of racial memories of events carried out long ago by the possible 'donors' of our excessive neurological endowments.

In the heady flying saucer days of the fifties, it is almost certain that those people who reported encounters with 'Venusians', who were described as tall, blonde-haired and blue-eyed beings, were projecting their own fears, concerns and worries over nuclear detonations and general planetary radiation levels into hypothetical and imaginary aliens that had come to Earth to voice their concern over our behaviour patterns. In the fifties these godlike figures appeared to certain people, such as the late George Adamski, in fairly remote areas and made their concern clear to these 'chosen' Earthlings,

Nowadays, the abductions which are alleged to occur involving hundreds of people, seem to be much more furtive, clandestine and insidious than the open, rather basic and naïve way in which the alien beings were supposed to have behaved in the fifties. Now, a much more believable picture seems to be emerging, with people re-living, in a sometimes fearful and traumatic manner, the experiences of their alleged abductions under hypnotic regression. Sometimes painful alien medical processes and tests were supposedly carried out on the victims by alien beings within the starships, the victims having been brought to the alien craft in a semi-subdued condition and taken to some kind of operating theatre within the craft.

In some cases, a kind of telepathic dialogue occurs and the same concern by the alien beings is 'voiced' over the question of human behaviour towards our planet and to each other. However this time, instead of just making clear their concern, it would appear that the aliens now have an active programme underway to do something positive about it, and these plans emerge in these hypnotic recall sessions.

Also, the emphasis is strongly made that genetic material removed from the male and female victims *is* being used to create quasi-human beings for some positive purpose by the abductors. In some cases the victims are actually shown the foetuses suspended in liquid and it is almost certain that, if these subjects who claim abduction are actual victims of real events, then these purposely reared beings will be used for some obviously quite important purpose later in their development. One day we ourselves will have isolated all the genes for various traits such as height, build, intellect and so forth. Perhaps *we* will one day create hybrid entities by genetic means, similar to the biblical 'giants', for specific purposes.

As previously stated, the alarming rate of advancement in Earthly genetic science makes it quite believable that beings only a few hundred years in advance of us would be quite capable of manipulating genes and constructing life forms without too much trouble. It has also been made clear by some sort of thought transference to their victims during these alleged abductions that the aliens see no possible way of appearing openly to mankind at the present time, as they seem to have predicted that our reaction to such revelations would be detrimental to themselves, and that humans would not be able to mentally handle it.

But more worryingly, it has been made clear to some of the victims that there is a clear intention to eventually breed enough beings to comfortably exist in our particular mixture of atmospheric gases, and that ultimately control of the Earth (if not a complete takeover) is the plan.

Is the depletion of the ozone layer part of that 'conditioning' of the Earthly environment? Perhaps stronger doses of ultraviolet radiation, while detrimental to us, may be quite beneficial to their metabolism. If their ancient ancestors were responsible for our creation long ago, we would have been genetically produced to function comfortably on this world, but may not have had a suitable enough metabolism for proper development on *their* world. The memory banks on their world would be vast and replete with information on the activities of their ancestors in outer space, as well as on their own discoveries and planetary analysis. If they now need another home, our planet may have come up in their computations and selection process as 'suitable subject to certain pre-conditioning'.

Some of the cases referred to in abduction stories and accounts in various books are extremely convincing accounts of alleged abductions of a wide range of age groups and professions. The well-documented and convincing Betty and Barney Hill case took place over thirty years ago, and the processes the victims were subjected to also involved genetic material being removed from them. If special beings have been created by hypothetical alien genetic experimentation, they would be mature adults by now. Are some of them already inserted and functioning in human affairs? If alien beings are present in our skies, their long and patient interest in humans may not mean their predecessors were responsible for our creation, but rather some other alien group who may be now residing in a remote part of the universe and busily constructing quasars as celestial navigation beacons for future space travellers like ourselves.

Many millions of years ago our ancient ancestors may have had an intellectual capacity similar to ours today. They would surely pass this way again, even if they are now manipulating some embryonic intelligences in other systems many light years away. The evidence of their visit may yet still be found in some ancient rock strata or undersea area, unless we have already found it in the aforementioned blue-green algae in ancient rocks or those ancient ribbed bootprints found in stone.

So, are we to conclude that we are the 'mighty men of renown' that the Bible speaks of as being the offspring of the gods who took unto themselves the daughters of men? Or are we just a freak, a mistake, an evolutionary misfit? We seem to sense, as humans, some higher presence. We seem to need it, which is why religions exist and why Superman movies and other celluloid heroes are so popular. Is it all part of the racial memories of looking up to our creator(s) as the monster did originally to his master the Baron Frankenstein? What put the idea in the lady author's head in the first place for such a thing, which in itself seems like a racial memory of a creation scenario? How is it that almost every earthly race has creation legends of men being 'made' by God or gods?

When I mentioned experimenting with dolphins, we must not forget selective breeding programmes with other creatures such as dogs and the many 'hybrid' types of plants produced by our manipulations. If we travelled to another planet and found docile hominid types slowly starving to death with foodstuff all around them, we may carry out some noble enterprise to assist them in their survival and go away feeling quite pleased with ourselves, and maybe even return to review our handiwork and further enhance it. It really could have all been done before.

Geneticists openly state that in the future we will be able to choose the sex, eye colour and maybe even the IQ level of our offspring. Also their height and athletic prowess, just as though we were picking out the model of car we wanted from a series of brochures. Perhaps other extraterrestrial races have long ago had this ability: they may have long isolated and now control the genes responsible for bodily and neurological traits and development.

The child prodigies and geniuses that appear from time to time may serve to reinforce the theory of extraterrestrial genes being present in our brain cells. Perhaps this is what they set out to do in the first place, but the genes responsible for it have disappeared into the great morass of the worldwide population explosion and only pop up once in a while in certain gifted individuals to remind us. If one goes back a couple of millennia, say to the ancient Greeks and less densely populated times, it would appear that great intellect was indeed quite commonplace and taken for granted, especially in theories, mathematical formulae and awareness of the Earthly environment and the cosmos. One would also have to include the ancient Egyptians, with their fantastic building, planning and design skills, unequalled thousands of years after their constructions and who strongly influenced Greek learning and achievements – all of which makes it puzzling to understanding the 'regression' of the Middle Ages, and a *decline* of science.

To repeat, we still have much of our brain material to stir into use. What will we be capable of in the future? We might well ask! We are already so far ahead of any other species on the planet with brain material still in reserve that the mind boggles. Having dwelt so long on the positive side of human capability and achievement, one can only hope that the cells will emerge to control our baser instincts.

Are we then, as we approach the end of the current millennium, on the eve of some very traumatic occurrences? *Has* Earth and its peoples been under intense observation and scrutiny at close quarters with all the alleged abductions? Is it almost the end of the benign and patient study and the final stage of some master plan? Is there a very serious reason for it all? Is their 'second coming' at hand? Are we created beings by unearthly intelligence? How would we receive such revelations? Would we be able to mentally handle it without severe social disorientation and cultural shock? How would 'they' go about it? Would they take over the entire media immediately after neutralising our defence capabilities with our heavy reliance on electronics in all our technology? This could be an obvious and possibly quite simple objective to them.

Would great interstellar craft slip quietly into Earth's orbit, or would they use the media first to forewarn us? Our fictional professor in the television debate may have been right when he said we undersell ourselves by discounting our past achievements, but what about our future achievements? Perhaps this fantastic lump of grey matter within our skulls, however we came by it, will take us back to deep space, from where we may have originated in the first place.

Epilogue

Although it is more comforting to dwell on the calculated life expectancy of our star and our scientific advancements which are 'taking us back one day from whence we came' etc., the reality may be quite different. We presume we haven't irreversibly destroyed the ozone layer, or that we won't have another ice age; that we needn't protect the planet with orbiting devices as an asteroid or comet will never again collide with the earth; that extraterrestrial intelligence does not exist in earth space and that *all* the abductees dreamt *all* of their alleged experiences and so we are not about to be subjected to extreme cultural shock and social disorientation. We also presume that human origins will be solved and all the necessary fossil links to prove the 'theory' (that has somehow turned into a 'dogma' without them) *will* one day be found.

We could alternatively fall back upon the more comforting thought that Genesis *is* right after all, and all the fantastic events involving 'angels' in the Bible that make it read in parts like a dramatic science fiction story, are from the imagination of the patriarchs such as Abraham and Moses, (who after all have not been historically proven to exist) and so extraterrestrial intelligence had no part in the creation of humanity.

Taking all the above points one by one: we may develop the ability to replenish the ozone layer; weather-control devices in orbit could alleviate an ice age; orbiting missiles under UN control, designed only to target a body smaller than the Earth, could be developed and placed in orbit. With regard to an extraterrestrial presence in earth space however, science *is* predisposed to the existence of extraterrestrials largely due to the calculation of there being a billion billion sunlike stars in the universe.

When we ask why we haven't so much as a shirt-button to show from them, we could also ask, would an intelligence capable of crossing interstellar space be dumb enough to leave evidence lying around everywhere unless they meant to? Is absence of evidence, evidence of absence?

There are nevertheless many incredible stories from many credible witnesses.

A recent television programme dealing with the paranormal, within which category the whole UFO enigma still seems firmly entrenched, showed a fantastic piece of film taken from space. It depicted an object closing in on the Space Shuttle. Suddenly, a laser beam shot up into space from a ground source and the object took immediate avoiding action by shearing off into space.

If this film is genuine and the laser beam was purposely fired, two things are possible. Firstly, those that fired it may have been simply attempting to eliminate just one of the 8,500 pieces of space hardware said to be rattling around up there that could pose a threat to the Space Shuttle, and if such technology has been developed, then this would be a sensible and peaceful use for it. However, if this was the case, those that fired the laser would have been utterly amazed at the object's avoiding action and if nothing hitherto had convinced the military of the reality of UFOs, then that event certainly would have done so.

Alternatively, the people who fired the laser may have been well aware that it was a UFO and may have done this before, when it was not recorded on film, and the rest of us have still not been told about it. In all the thousands of UFO reports we would be hard-pressed to find a single case of planned aggression against humanity, yet *we* could be trying to kill *them*. Have these people the right to put the entire human race in jeopardy? What is going on out there in all these remote desert locations such as 'Area 51'?

A $10,000 fine is threatened against 'rubberneckers' or investigative journalists and a much more serious fate may befall a foreign secret agent or spy. (Perhaps the authorities should remember that Sodom and Gomorrah were in a remote desert location.) Their actions in any possible alien annihilation attempts could be a very profound turning point for humankind. Is there an alien 'final solution' to the human problem? Is this what Nostramadus foresaw in his 'Great King' coming from the skies 'when Mars (War) reigns supreme'?

I suggested in the book that SETI (the Search for Extraterrestrial Intelligence) might be an expensive ruse to make us think that, 'As we are still looking, we cannot have the evidence of alien existence in our laps, can we?'

Consider the following interesting facts. The brother of Frank Drake (the scientist heavily involved in SETI) works in one of those remote locations, the Los Alamos National Laboratory, New Mexico. Frank Drake appears to be preparing the world for something quite profound. Have the authorities now decided to abandon the cover-up they have so long been accused of? Are they now trying to prepare the world for a major event? These amazing statements were made by Frank Drake in the book *Is Anyone Out There?* (Simon & Schuster).

> I am telling my story because I see a pressing need to prepare thinking adults for the outcome of the present search activity, that is, the imminent detection of signals from an ET civilisation. This discovery, which I fully expect to witness before the year 2000 (2 years away) will profoundly change the world.

If this alien intelligence does actually exist already in earth space, they must be controlled from their point of origin, which in this work I have called 'Planet Heaven'. The abduction policy must be directed from there. Since the earliest convincing abductions such as that of Betty and Barney Hill, over thirty years have gone by. There could be many possible alien hybrid creations now in existence, going through an intensive preparation process for 'The Second Coming'. What will happen to us lesser mortals we can only guess, but at the very least we should see that the purposeful firing of laser beams must be extremely ill-advised, and whatever the authorities are up to in those remote locations, they may reflect on the possibility that they may be the first to go if the policy that seemed to be condoned in biblical times by 'Planet Heaven' is reintroduced. No one would wish that no attempts be made to defend humanity from a science-fiction type of 'Alien Invasion' scenario, but as said, there doesn't seem to be any evidence of aggression or evil intent from extraterrestrial intelligence. Humans however, have *never* lost the proclivity for naked aggression. Even if there are no alien life forms anywhere near earth, we still go in search of God or the 'gods' by peering ever closer at the tiniest particles of all matter, trying to discover the secrets of life itself. The geneticists, with their 'super sequencer' machines and more advanced models that will follow them, will eventually isolate (and manipulate) all the genes responsible for human traits, behaviour and bodily ailments. *We* will then become the creators.

It wasn't surprising that the Church elders feared science and called it 'Satanic'. They knew it wasn't a passing craze and that the ultimate end would be similar to Lucifer's challenge to God (or the gods), but what they failed to realise was that God (or the gods) *provided* the means for such a confrontation in the first place – the human brain.

After 50 million years of evolution, apes are still light years away from the ability even to count their fingers, but humans seem to have been given a wonderful capacity for mathematics along with other fantastic over-endowments of the human brain, and without mathematics to work out all the precise calculations for space flight and re-entry 'burns', our space explorations would have been utterly impossible. Dr Max H. Fundt, who inspired me to write *The Human Question* along with the NASA communications expert Otto Binder, said in the book *Mankind Child of the Stars* (Coronet):

> What applies to human space endeavours applies also to extraterrestrials. If the earth was once visited by extraterrestrials, these visitors must have been well versed in mathematics, that is why I consider our capacity for mathematics to be an indication that we are not only of earthly origin.

If humanity does manage to isolate the genes responsible for our base negative qualities, would we necessarily eliminate them? The possibility exists that we would not utilise the ability to cure these negative human traits. After some discussion someone would be bound to say, 'Whoah! Hold it. Do we want to become a nation of wimps? We may need these qualities to defend ourselves and the planet against future alien invaders.'

This of course is based on the assumption that the continuing development of civilised qualities may not necessarily have happened in other beings by the time they had the means of interstellar travel. We don't throw Christians to the lions anymore, so there are pointers indicating *some* natural loss of these traits. Nevertheless, the process seems too slow in comparison to our scientific advancements towards interstellar travel, and so we assume hypothetical alien invaders will have the same aggressive qualities. Moreover, we may actually develop android or cybernetic beings as disposable beings with these genes or aggressive qualities actually *enhanced* in their brains to do our dirty work for us.

What's more, these factors may have already been evaluated and computed by the hypothetical descendants of our original creators, who are now (possibly) having to dodge our purposely fired laser beams and attempts to destroy other life forms. This is a very worrisome possibility that may not bode well for the future (or lack of it) of the human race.

With regard to human creation, many *would* like to ignore their more logical side and accept the biblical Genesis account, but cannot do so. Is there a connection between the 'beings of light' mentioned in NDEs or near death experiences and the same type of beings in reported extraterrestrial encounters? There is a fantastic account in *The NDE Experience* (Routledge) where Mellen Thomas

Benedict 'died' for an hour and a half in 1982 then came back from his NDE and promptly recovered not only his life functions but also from terminal cancer. The events of his NDE were more fantastic than David Bowman's trip through the Stargate in the movie *2001: A Space Odyssey*.

In *The NDE Experience*, a being of light communicated with him telepathically and stated that, 'humans were given the power to heal themselves before the beginning of the world.'

This last phrase entirely agrees with the theory that extraterrestrial intelligences could have bequeathed part of their intellect and brain power that evolved before the solar system was born, to humanity. 'Beings of light' have appeared to humans in places that have become holy places visited by people seeking cures. Many have been cured, who had the necessary applied mind power called 'faith' to achieve it.

Regarding our struggle to find proof of simian ancestry, I have stated that there is a kind of 'desperation' to find it, for the simple reason that having rejected a biblical creation where else could one turn? Where else, but to the theory of extraterrestrial 'genetic' creation? I mentioned the amazing potential for error that exists in digging up old bones and the work *Human Remains* (The British Museum Press, 1994) bears this out and says:

> Isolated fragments of bones and teeth can easily be misidentified, fragmentary human bones are sometimes being confused with those of large carnivores such as bears... fragments of the limb bones of large birds, such as swans or geese, are also mistaken for human remains.

If the amazing theory of alien genetic creation is right, the descendants of the creators, acting so furtively *now* in earth space did not fear humans in biblical times, nor had any reason to fear human weaponry or aggression (we are now firing laser beams at them) and moved among the patriarchs carrying out their various operations which included mass annihilation of humans.

I have highlighted these incredible events in *The Angels of Abraham*. One amazing event had the angels using some type of 'neutron bomb' or 'biological' method, annihilating 185,000 Assyrians in their sleep.

When the Hebrews awoke in their desert location, their enemies 'were all dead corpses'; they didn't have to lift a finger in the operation (2 Kings 19:35–36). The British and Foreign Bible Society clearly attributes the writings of Genesis and Exodus to Moses, and Genesis relates the life and experiences of Abraham. The Old Testament is based on the Hebrew Bible of the Jews with the New Testament being the Christian documents.

Certain books dealing with biblical matters state that the Hebrew Bible was only begun in AD 70, after the destruction of the Temple by the Romans and was not completed in a single stabilised form until 1,000 years later. But the Hebrews believed in older, verbally handed-down accounts and that God did give Moses, personally, the data for the Ten Commandments and his other writings.

Races such as the Egyptians took a lot of trouble to meticulously record their events. Other races in the area must also have done so, on skins and so forth, which seems to have been born out by the finding of the Dead Sea Scrolls. Nevertheless, the possibility exists that all human creation legends which agree that men were 'made' by their

God or gods are based on the amazing possibility of an extraterrestrial genetic experiment with the primates that could have occurred up to 50,000 years ago. For all that's been said in the work regarding the possible annihilation of masses of humanity in floods and bombing iniquitous cities, described in the biblical accounts, when eliminating humans considered to be genetic 'failures' or 'freaks', how would the descendants of the possible creators view the entire human question today?

Ever since the first 'experiment' when Cain 'malfunctioned' and slew his brother out of jealousy, humans are still actively slaughtering each other. They have now seemingly turned on the descendants of our (hypothetical) creators, by shocking them out of the sky and attempting to kill them. What is the purpose of the armies of 'hybrid creations' that appear to be under production according to convincing data extracted in hypnotic regression sessions with all those alleged abductees? Will they become the supreme genetic creations destined to inherit the Earth? What will happen to the rest of us?

In all the millennia that have passed since the possible initial creation, humans are advancing rapidly in cosmic exploration, planning and Star Wars weaponry, which we now appear to have turned on them. We don't appear to have advanced at all in reducing our tendency to kill each other. Have we therefore made our life form redundant and surplus to requirements by our own actions?

Reference:

When Charles Lyell finally came round, in his Geological Evidence Antiquity of Man [1863] his book sold 4,000 copies in the first few weeks, two new editions appeared in the same year.

Since then, as we shall see in Chapter 1 ancient stones tools have been found all over the world, and their distribution and variation enable us to recreate a great deal about our distant past and the first ideas and thoughts of ancient humankind. In the century and a half since Prestwich and Evans confirmed de Perthe's discoveries, the dating of the original manufacture of stone tools has been pushed back further and further, to the point where this book properly starts: the Gona river in Ethiopia 2.7 million years ago.

prologue Ideas: A History From Fire to Freud.
Peter Watson.